T0297852

Series on Analysis, Applications and Computation – Vol. 4

ISAAC

Nonlinear Waves
An Introduction

Series on Analysis, Applications and Computation

Series Editors: Heinrich G W Begehr *(Freie Univ. Berlin, Germany)*
Robert Pertsch Gilbert *(Univ. Delaware, USA)*
M. W. Wong *(York Univ., Canada)*

Advisory Board Members:
Mikhail S Agranovich *(Moscow Inst. of Elec. & Math., Russia)*,
Ryuichi Ashino *(Osaka Kyoiku Univ., Japan)*,
Alain Bourgeat *(Univ. de Lyon, France)*,
Victor Burenkov *(Cardiff Univ., UK)*,
Jinyuan Du *(Wuhan Univ., China)*,
Antonio Fasano *(Univ. di Firenez, Italy)*,
Massimo Lanza de Cristoforis *(Univ. di Padova, Italy)*,
Bert-Wolfgang Schulze *(Univ. Potsdam, Germany)*,
Masahiro Yamamoto *(Univ. of Tokyo, Japan)* &
Armand Wirgin *(CNRS-Marseille, France)*

Series on Analysis, Applications and Computation – Vol. 4

ISAAC

Nonlinear Waves
An Introduction

∘ Petar Popivanov
Bulgarian Academy of Sciences, Bulgaria

∘ Angela Slavova
Bulgarian Academy of Sciences, Bulgaria

World Scientific

NEW JERSEY • LONDON • SINGAPORE • BEIJING • SHANGHAI • HONG KONG • TAIPEI • CHENNAI

Published by

World Scientific Publishing Co. Pte. Ltd.

5 Toh Tuck Link, Singapore 596224

USA office: 27 Warren Street, Suite 401-402, Hackensack, NJ 07601

UK office: 57 Shelton Street, Covent Garden, London WC2H 9HE

British Library Cataloguing-in-Publication Data
A catalogue record for this book is available from the British Library.

NONLINEAR WAVES: AN INTRODUCTION
Series on Analysis, Applications and Computaitons — Vol. 4

ISBN-13 978-981-4322-12-6

Printed in Singapore.

Preface

This book deals with several equations of Mathematical Physics as Korteveg-de Vries (KdV) equation and its different modifications (shortly mKdV), the Camassa-Holm equation and its generalizations, the nonlinear hyperbolic equation describing the vibrations of a chain of particles interconnected by springs, the viscoelastic generalization of Burger's equation, the sin-Gordon equation, the Hunter-Saxton equation and others. They originate from Physics — for example Camassa-Holm equation is modeling the propagation of unidirectional irrotational shallow water waves over a flat bad — but we proposed here their investigation via purely mathematical methods in the frames of University courses (Lebesgue integral and its properties, Fourier transform, Schwartz distributions and the corresponding functional spaces, Ordinary Differential Equations (ODE), method of the characteristics, etc.). Therefore, the book is addressed to a broader audience including graduate students, Ph.D. students, mathematicians, physicists, engineers and specialists in the domain of Partial Differential Equations (PDE) and their applications. Certainly, there are monographs on the subject based on complicated and difficult methods that make the readers acceptance hard — especially for beginners and non-specialists on PDE. We give below a short description of this book.

The first two chapters are devoted to the travelling wave solutions of the above mentioned different type of equations of Mathematical Physics. The corresponding solutions (geometrically their profiles) may have different kind of singularities — peaks, cusps, compactly supported and are called peakons, cuspons, compactons. Certainly, solitons, kinks and periodic solutions can appear and we have found the asymptotic behavior of the solutions near the singular points.

Chapter III is central for the book. At first for a sufficiently large class of third order nonlinear PDE of 2 independent variables we propose two classification theorems for the corresponding travelling wave solutions. Moreover, the solutions can develop different isolated singularities and we study them very precisely. Then we give a short survey on the theory of elliptic functions, where the Jacobi functions $sn(x, k)$, $cn(x, k)$, $dn(x, k)$ and the elliptic integrals (Legendre functions) of first, second and third kind $F(\varphi, k)$, $E(\varphi, k)$, $\Pi(\varphi, n, k)$ are not only defined but their main properties are formulated as well. This way we are able to express explicitly the travelling wave solutions of many equations of Mathematical Physics by the above mentioned elliptic functions. As far as we know the explicit form of these solutions is highly appreciated by physicists. Interesting properties of special type solutions (non travelling waves) of the sin-Gordon equation are discussed too. Then the so called fluxons, antifluxons and breathons can appear and we investigate the interaction fluxon-antifluxon and fluxon-fluxon after their collision.

We point out that the travelling wave solutions are not very special objects. In fact, in Chapter IV we study the stability (orbital stability) of the periodic solutions — travelling wave type — for some mKdV type equations. The orbital stability results are valid for the solutions having the same minimal period of the initial data. In the proof of the stability results a crucial role is played by the Jacobi elliptic functions and precise theorems from the spectral theory for second order ODE. To avoid the technicalities, to stress the main ideas and to simplify the proofs we propose a short sketch only. Thus, delicate results from the harmonic analysis and the spectral theory are not given here. Instead of, some comments and illustrative examples are proposed to the reader.

In Chapter V the interaction of two peakons of special kind satisfying in weak sense the Camassa-Holm equation is studied. Our results are based on the jump formula for the Schwartz distributions and the explicit formulas for the solutions of a Hamiltonian system. In several papers the interaction of $n \geq 2$ peakons is investigated but we think that the case $n = 2$ is more transparent and easy to understand.

In Chapter VI we apply the method of characteristics to the solvability of the Cauchy problem for second and third order Hunter-Saxton equation. This equation appears in chemistry. At first, classical solutions are constructed and the unicity problem is discussed. Moreover, we find global in time solutions as well as we estimate the life span of the solution when it blows up in finite time. The important and delicate problem for the existence of weak solutions is stated and discussed in Section 2 of Chapter VI.

To avoid the technical proof we formulate only the main existence result and give some examples.

We propose in Chapter VII a multicomponent generalization of the Camassa-Holm equation and construct peakon type solutions of special form. A detailed geometrical study of the corresponding solutions is given including their asymptotic behavior for $t \to \infty$. The integral surfaces here obtained are partially ruled.

Chapter VIII is the second central chapter of this book. At first we discuss the Riemann problem for a special system of conservation law (first order quasilinear hyperbolic system in the plane). It turns out that it generates the so called δ-shock. We give the corresponding definition of a generalized δ- shock wave type solution and we show that the weak solution of the above mentioned Riemann problem satisfies the assumptions of that definition.

In the second paragraph of the same Chapter we study the weak continuous solutions for the scalar conservation laws. We propose here all the necessary details for proving the existence of a continuous solution. Moreover, it turns out to be Hölder one and differentiable almost everywhere in $\{t > 0\}$.

The last Chapter IX studies a first order non strictly hyperbolic system in $\mathbf{R}^3_{t,x}$ having initial data of the type "finite jumps". It is proved that due to the interaction of two piecewise smooth waves logarithmic singularities appear. The above described effect does not arise in $\mathbf{R}^2_{t,x}$. We point out that the interaction is realized along a straight line not contained in a space like manifold. The proofs here rely on well known results from the microlocal analysis.

As it concerns numerical methods in the case of periodic waves we apply CNN (Cellular Neural Network) approach. Short introduction on this subject is proposed in Chapter I.

Our examples of nonlinear waves are illustrated by about 50 figures. The results from Chapters I, II and partially from Chapters III and IX were recently obtained and published by the authors. The rest part is due to other mathematicians but we hope that it is incorporated in a natural way in this book. Having in mind that this is an introduction to the theory of nonlinear waves we have proposed a list of papers and books in the references that can enlarge the readers knowledge on the subject.

Petar Popivanov
Angela Slavova

Contents

Chapter 1

Compact travelling waves and peakon solutions and their Cellular Neural Network realization

1.1 Introduction and formulation of the main results

1. This Chapter deals with travelling wave solutions of several classes of partial differential equations (PDE) of mathematical physics having different type of singularities.

At first we define a compact travelling wave (see [103], [101]) as a travelling wave having compact support beyond which it vanishes identically. We remind of the reader that according to [101] a compact wave is a robust solitary wave with compact support, while the compacton is a compact wave that preserves its shape after the interaction with another compacton. The problem of interaction of compact travelling waves is rather delicate as they are unstable in some sense. Because of this fact we shall concentrate on the existence of compact travelling waves only. It is classical that the soliton is a special solitary travelling wave that after a collision with another solution eventually emerges unscathed ([119], [16]). Solitons appear in the propagation of water waves or waves along a mass-spring chain. Travelling wave solutions of different classes of PDE are studied in many papers. We shall mention several of them only as they are closely connected with the content of this chapter: [23], [24], [29], [93], [121], [97], [101].

2. In order to introduce the notion of peakon (travelling wave with peaks) we shall define the peak type singularity of the graph of the function $\lambda = \varphi(s) \in C^0$ at the point $P_0 = (s_0, \lambda_0 = \varphi(s_0))$. We assume that φ has a local maximum (local minimum) at s_0, and that there exist left and right derivatives of φ at s_0, including the case of $|\varphi'(s_0 \pm 0)| = \infty$. Therefore, either the corresponding tangent lines to the curve at P_0 are not vertical

1

and not coinciding (for $\varphi'(s_0 + 0) \neq \varphi'(s_0 - 0)$), or the curve possesses a vertical tangent at P_0. The curve has a cusp type singularity at P_0 if, say, $(\lambda - \lambda_0)^{2k+1} = \pm(s - s_0)^{2l}$; $k, l \in \mathbf{N}$, $2k + 1 > 2l$.

To be more precise, we shall say that the wave is of the type peakon-cuspon if at least in one point Q_0 the function $\varphi(s)$ develops cusp type singularity. Assume that $supp\ \varphi$ is compact, say $\varphi \leq 0$ and $supp\ \varphi = [a, b]$. If φ is smooth at the end points a, b of the compact the travelling wave is called smooth compactly supported wave. Suppose that $\varphi'(a + 0) \neq 0$ and $\varphi'(b - 0) = \infty$. Then we shall speak about peakon-cuspon compactly supported wave etc. Briefly, without confusion, we shall say in the previous case that the solution is compacton-peakon, compacton-cuspon etc.

Definition 1.1. We shall say that the travelling wave $u = \varphi(x - ct)$, $c > 0$ is a peakon wave, briefly: peakon if the graph $\Gamma = \{s, \varphi(s)\}$ of the function $\lambda = \varphi(s)$ possesses at least one peak type singularity.

We shall deal with continuous solutions which are smooth except to a finite set of points (peaks, end points of the compact support).

Interesting examples of PDE having compact travelling wave solutions and/or peakons are given by the Camassa-Holm or generalized Camassa-Holm equations [29], the generalized Korteveg-deVries (KdV) equation [119], [101]–[103], the nonlinear PDE describing the vibrations of a chain of particles interconnected by springs [101]. Below we shall not discuss the physical interpretation of the above mentioned classes of PDE and the corresponding travelling wave solutions (compactons, peakons, solitons, etc.).

3. Define now the following generalization of the Camassa-Holm equation:

$$u_t + K(u^m)_x - (u^n)_{xxt} = \left[\frac{((u^n)_x)^2}{2} + u^n(u^n)_{xx}\right]_x, \qquad (1.1)$$

where $K = const > 0$ and $m, n \in \mathbf{N}$. The standard Camassa-Holm equation is obtained for $n = 1$, $m = 2$, $K = 3/2$.

We point out that in many cases the assumption that n, m are positive integers can be omitted ($n, m \in \mathbf{Z}$).

Put $u = \varphi(x - ct)$, $c = const > 0$, $\varphi(0) = \bar{c}$ in (1.1), i.e. we are looking for travelling wave solutions of (1.1). Usually, it is supposed that $\varphi(\infty) = \varphi'(\infty) = \varphi''(\infty) = 0$. In other words, infinite tails of the waves are admitted. For us it is interesting and we suppose that it is interesting from physical point of view too to construct nontrivial compact travelling waves satisfying (1.1), peakons etc.

To simplify the formulation of Theorem 1.1 we propose the following assumption (A):

(A) $c^{\frac{n-m+1}{n}} > \frac{n+1}{m+n}K$, where $\varphi(0) = \bar{c} = c^{\frac{1}{n}}$.

This is the first result of this Chapter.

Theorem 1.1. *1) Let either $m = 2$, $n = 1$, $K = 3/2$ (classical Camassa-Holm equation) or $m = -1$, $n = -2$, $K = 3$. Then the equation (1.1) possesses a peakon, but not compact travelling wave solution.*

2) Let $n = 1$, $m = 2$, $0 < K < 3/2$. Then (1.1) has a peakon-cuspon type solution forming a cusp type singularity at $(0, c)$, i.e. $\varphi(s)$ has a cusp type singularity at $s = 0$, $\varphi(0) = c$.

3) Let $n > 3$, $m \geq 1$, and the condition (A) holds.

Then (1.1) possesses a compact travelling wave-peakon-cuspon type solution developing a cusp type singularity at the point $(0, c^{1/n})$, i.e. $\varphi(s) \sim c^{1/n} - const|s|^{2/3}$, $s \to 0$, $const > 0$.

The solutions constructed in Theorem 1.1 are even, i.e. $\varphi(-s) = \varphi(s)$. The case $1 \leq n \leq 3$, $m \geq 1$ can be treated in a similar way as in Theorem 1.1 and is left to the reader.

This is a generalization of the KdV equation

$$u_t + [a(u)]_x + [b(u)(u^n)_{xx}]_x = 0, \qquad (1.2)$$

where $a(u)$, $b(u)$ are sufficiently smooth functions (for the sake of simplicity we can assume that $a, b \in C^\infty$) and such that

(B) $b(u) \neq 0, \forall u, a(u) = u^m + O(u^{m+1}), u \to +0,$
$a(u) > 0$ for $u > 0$, $n \geq 4$, $m \geq n - 2 \geq 2$; $m, n \in \mathbf{N}$.

Without loss of generality we shall suppose that $b(u) > 0$ everywhere.

The case $a(u) = u^m$, $b(u) = 1$, $m > 0$, $1 < n \leq 3$, is considered in [103].

Introduce now the functions:

$$A(z) = \int_0^z \frac{\mu^{1/n} d\mu}{b(\mu^{1/n})} = n \int_0^{z^{\frac{1}{n}}} \frac{\lambda^n d\lambda}{b(\lambda)}, z \geq 0 \qquad (1.3)$$

$$B(z) = \int_0^z \frac{a(\mu^{1/n}) d\mu}{b(\mu^{1/n})} = n \int_0^{z^{\frac{1}{n}}} \lambda^{n-1} \frac{a(\lambda)}{b(\lambda)} d\lambda, z \geq 0,$$

$$A_1(\varphi) = c \frac{A(\varphi^n)}{n^2 \varphi^{2n-2}}, B_1 = \frac{B(\varphi^n)}{n^2 \varphi^{2n-2}}, \varphi \geq 0 \qquad (1.4)$$

and assume that the following condition is satisfied:

(C) $cA(z) > B(z) > 0, 0 < z < D$ for some D and $cA(D) = B(D)$,
$cA'(D) \neq B'(D)$, i.e. $cD^{1/n} \neq a(D^{1/n})$.

We point out that for every $c > 0$ there exists $D(c) > 0$ and such that $cA(z) > B(z)$ for $0 < z < D$.

Evidently, according to (1.3) $A(0) = B(0) = 0$ and $A(z) > 0$, $B(z) > 0$ if $z > 0$.

This is the second result of Chapter I.

Theorem 1.2. *Under the assumptions (B), (C) the equation (1.2) possesses a nontrivial compact travelling wave solution.*

Possible generalizations of the results proved in Theorems 1.1, 1.2 and applications to the nonlinear PDE modelling the vibrations of chain of particles interconnected by springs are given in Section 3.

1.2   Proof of Theorem 1.1

1. We are looking for a travelling wave solution of the Camassa-Holm type equation (1), i.e. $u = \varphi(x - ct)$, $c = const > 0$. Put $s = x - ct$ and suppose that $\varphi(s)$ is even function. In the case of peakons we assume that $\varphi(\infty) = \varphi'(\infty) = \varphi''(\infty) = 0$, while in the case of compact travelling waves or compact travelling waves-peakons the function $\varphi(s)$ is compactly supported. Substituting $\varphi(x - ct)$ in (1.1) and denoting $\varphi(0) = \bar{c} > 0$, we get

$$c(\varphi^n)''' - c\varphi' + K(\varphi^m)' = \left[\frac{((\varphi^n)')^2}{2} + \varphi^n(\varphi^n)'' \right]', \qquad (1.5)$$

i.e.

$$c(\varphi^n)'' - c\varphi + K\varphi^m = \frac{((\varphi^n)')^2}{2} + \varphi^n(\varphi^n)'' + C_1. \qquad (1.6)$$

To simplify the things we put $C_1 = 0$ and suppose that $\varphi \geq 0$ everywhere. With $z(s) = \varphi^n(s) \iff \varphi(s) = z^{1/n}(s)$ we have:

$$-cz^{\frac{1}{n}}(s) + Kz^{\frac{m}{n}}(s) + cz''(s) = \frac{(z')^2}{2} + zz''. \qquad (1.7)$$

Multiplying both sides of (1.7) by z' and integrating again with respect to the variable s we obtain:

$$-c\frac{z^{1+\frac{1}{n}}}{1 + \frac{1}{n}} + K\frac{z^{\frac{m}{n}+1}}{\frac{m}{n} + 1} + \frac{c}{2}(z')^2 = \frac{z(z')^2}{2} + C_2. \qquad (1.8)$$

Assuming $C_2 = 0$, we rewrite (1.8) in the form:

$$(z')^2(c - z) = c_1 z^{1+\frac{1}{n}} - K_1 z^{\frac{m}{n}+1}, \tag{1.9}$$

where $c_1 = \frac{2c}{1+\frac{1}{n}}$, $K_1 = \frac{2K}{\frac{m}{n}+1}$.

Going back to the function $\varphi = z^{\frac{1}{n}}$ we have:

$$n^2\varphi^{2(n-1)}(\varphi')^2(c - \varphi^n) = c_1\varphi^{n+1} - K_1\varphi^{m+n}, \varphi(0) = \bar{c} > 0. \tag{1.10}$$

So we get,

$$n^2\varphi^{2(n-1)}\left[(\varphi')^2(c - \varphi^n) - \frac{c_1}{n^2}\varphi^{3-n} + \frac{K_1}{n^2}\varphi^{m+2-n}\right] = 0. \tag{1.11}$$

Introducing new constants $c_2 = \frac{c_1}{n^2} = \frac{2c}{n^2+n}$, $K_2 = \frac{K_1}{n^2} = \frac{2K}{n(m+n)}$ we reduce our problem to the solvability of the following ODE with separate variables:

$$(\varphi')^2(c - \varphi^n) = c_2\varphi^{3-n} - K_2\varphi^{m+2-n}, \tag{1.12}$$

i.e.

$$(\varphi')^2(c - \varphi^n) = \begin{cases} \varphi^{3-n}(c_2 - K_2\varphi^{m-1}) \text{ or} \\ \varphi^{m+2-n}(-K_2 + c_2\varphi^{1-m}), \end{cases} \tag{1.13}$$

$\varphi(0) = \bar{c}$.

In the case of classical Camassa-Holm equation $m = 2$, $n = 1$ we assume that $\bar{c} = c$, $c = c_2$, $K_2 = 1$, i.e. $K = \frac{3}{2}$ and (1.12) takes the form $(\varphi')^2 = \varphi^2$ if $\varphi \not\equiv c$.

Therefore, the classical Camassa-Holm equation possesses a peakon, but not compact travelling wave solution $\varphi(s) = ce^{-|s|}$ (we take $\varphi' = -\varphi$ for $s \geq 0$ and $\varphi' = \varphi$ for $s \leq 0$). This is well known fact.

Suppose now that $n = m - 1$, $c = c_2$, $K_2 = 1$. Then $m = n + 1$, implies that $n_1 = 1$, $m_1 = 2$, $K = \frac{3}{2}$ or $n_2 = -2$, $m_2 = -1$, $K = 3$. In the second case (1.13) takes the form $(\varphi')^2 = \varphi^5$, $\varphi(0) = \bar{c}$ and easy computations give us the following peakon solution: $\varphi(s) = (\frac{3}{2}|s| + \bar{c}^{-3/2})^{-2/3}$.

Consider again the case $n = 1$, $m = 2$ under the additional restriction $K_2 < 1$, i.e. $0 < K < \frac{3}{2}$ and $c = \bar{c} = c_2 = \varphi(0)$. Then $(\varphi')^2(c - \varphi) = \varphi^2(c - K_2\varphi)$, $c - K_2\varphi > c - \varphi \geq 0$, $\varphi \geq 0$. The corresponding peakon solution is given implicitly by the following formula after the computation of some integrals

$$\frac{\left(1 - \sqrt{\frac{c-\varphi}{c-K_2\varphi}}\right)\left(1 + \sqrt{\frac{c-\varphi}{\frac{c}{K_2}-\varphi}}\right)^{\sqrt{1/K_2}}}{\left(1 + \sqrt{\frac{c-\varphi}{c-K_2\varphi}}\right)\left(1 - \sqrt{\frac{c-\varphi}{\frac{c}{K_2}-\varphi}}\right)^{\sqrt{1/K_2}}} = e^{-|s|}. \tag{1.14}$$

There is a big difference among the peakons obtained in the previous cases and (1.14). In fact, the graph of the solution φ in the cases $n_1 = 1$, $m_1 = 2$, $K = \frac{3}{2}$; $n_2 = -2$, $m_2 = -1$, $K = 3$ forms an angle with positive opening at the point $(0, \bar{c})$, while (1.14) forms a cusp type singularity at $(0, \bar{c})$.

2. Below we shall deal with the case $n > 3$, $m \geq 1$ and we shall find solutions of (1.13) of compact travelling wave — peakon type. Our main assumption (A) can be rewritten as:

(A') $\varphi(0) = \bar{c} = c^{1/n}$,

$$c_2 - K_2 c^{\frac{m-1}{n}} > 0.$$

At first we shall construct the solution for $s \geq 0$. To do this we consider the ODE

$$\varphi' = -\frac{\varphi^{\frac{3-n}{2}}}{\sqrt{c - \varphi^n}}\sqrt{c_2 - K_2\varphi^{m-1}}, \varphi(0) = c^{1/n} > 0.$$

Suppose that $0 \leq \varphi \leq c^{1/n}$. Then the function

$$H(\varphi) = \int_\varphi^{c^{1/n}} \frac{\sqrt{c - \lambda^n}\lambda^{\frac{n-3}{2}}}{\sqrt{c_2 - K_2\lambda^{m-1}}}d\lambda \tag{1.15}$$

is well defined according to (A'), is monotonically decreasing in the interval $\varphi \in [0, c^{1/n}]$ and is a bijection onto the interval $s \in [0, d]$, where $d = \int_0^{c^{1/n}} \frac{\sqrt{c - \lambda^n}\lambda^{\frac{n-3}{2}}}{\sqrt{c_2 - K_2\lambda^{m-1}}}d\lambda$; $H(0) = d$, $H(c^{1/n}) = 0$. Moreover, $H : (0, c^{1/n}) \to (0, d)$ is a diffeomorphism, $H'(+0) = 0$, $H'(c^{1/n}) = 0$. The solution we are looking for is $\varphi = H^{-1}(s)$, $s \geq 0$, $\varphi(0) = c^{1/n}$, $\varphi(d) = 0$, $\varphi'(+0) = -\infty$, $\varphi'(d) = -\infty$. In the case $s \leq 0$ we consider the equation

$$(c - \varphi^n)(\varphi')^2 = \varphi^{\frac{3-n}{2}}\sqrt{c_2 - K_2\varphi^{m-1}}, \varphi(0) = c^{1/n}$$

and we construct φ by the formula $-H(\varphi(s)) = s \leq 0$. Thus,

$$H(\varphi) = |s|, |s| \leq d \tag{1.16}$$

and obviously, $\varphi(s)$ is even function.

To complete our study we shall find the behaviour of $H^{-1}(s)$, $0 \leq |s| \leq d$ near to the singular points $s = 0$, $s = d$.

If s describes some neighborhood of the origin then φ describes some neighborhood (one-sided) of $c^{1/n}$ and consequently $H(\varphi)$ is equivalent to const. $\int_\varphi^{c^{1/n}} \sqrt{c^{1/n} - \lambda}d\lambda$ as $a^n - b^n \approx (a - b)(a^{n-1} + b^{n-1})$ if $a \geq 0$, $b \geq 0$. So we have that near $s = 0 : |s| \approx const(c^{1/n} - \varphi)^{3/2}$, i.e. $c^{1/n} - \varphi \approx const|s|^{2/3}$, the const being positive. The graph of our solution develops

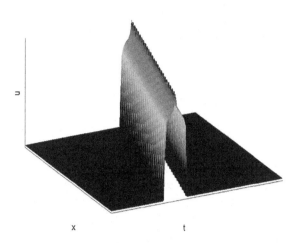

Fig. 1.1 Peakon wave.

a cusp type singularity at $(0, c^{1/n})$. Having in mind the fact that $|s| = \int_{\varphi}^{0}(...) + d = H(\varphi)$ we can easily see that near $\varphi = 0 : d - |s| \approx const\, \varphi^{\frac{n-1}{2}}$, i.e. $\varphi \approx const(d - |s|)^{\frac{2}{n-1}}$ near $|s| = d$, $|s| < d$.

Geometrically we have Fig. 1.1:

We continue $\varphi(s)$ as 0 for $|s| \geq d$. The solution φ has a singularity at $s = 0$ in the sense that $\varphi' \approx -const|s|^{-1/3}$ near 0. The left hand side of the equation (1.10) does not have singularity at $|s| = 0$ because $(\varphi')^2(c - \varphi^n) \approx const|s|^{-2/3}[c - (c + O(|s|^{2/3}))]$, $|s| \to 0$. There are no difficulties to verify that the left hand side of (1.10) does not have singularity at $|s| = d$ too because $\varphi(d) = 0$, $\varphi^{n-1}\varphi' \approx const(d - |s|)^2(d - |s|)^{\frac{3-n}{n-1}} = const(d - |s|)^{\frac{n+1}{n-1}}$, for $|s| \to d$. In other words, the compacton-peakon constructed by the formula (1.16), $\varphi = 0$ for $|s| \geq d$ satisfies our equation (1.10), respectively (1.12).

Remark 1.1. Suppose that $n = 3$, $m > 1$. Then according to (1.15), (A') we have that for $0 \leq s \leq d$, $H(s)$ is strictly monotonically decreasing, $0 \leq H(\varphi) \leq d$, $H'(+0) = -\sqrt{\frac{c}{c_2}} < 0$, $H'(c^{1/3}) = 0$. Therefore, the graph of $\varphi(s) = H^{-1}(s)$ (see Fig. 1.2) is different from the graph, proposed on Fig. 1.1 (Theorem 1.1, 3).

The singularity of φ is cusp type at $(0, c^{1/3})$, while at $(|d|, 0)$ the slope of the tangents is finite.

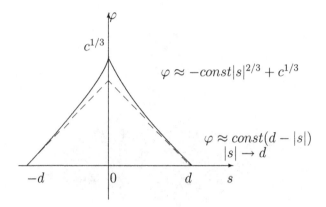

Fig. 1.2

Thus everything is proved.

1.3   Proof of Theorem 1.2

Searching for a travelling wave solution of (1.2) we put $u(t,x) = \varphi(x - ct)$, $s = x - ct$, $c = const > 0$. Then (1.2) takes the form:

$$-c\varphi' + \frac{d}{ds}a(\varphi(s)) + \frac{d}{ds}[b(\varphi)(\varphi^n)''] = 0.$$

Put $z(s) = \varphi^n(s)$, $\varphi \geq 0$, i.e. $z(s) \geq 0$.

Therefore,

$$-c\varphi(s) + a(\varphi(s)) + [b(\varphi)(\varphi^n)''] = C_1 = const.$$

Taking $C_1 = 0$ and multiplying both sides of the previous equation by $z'(s)$ we get:

$$-cz^{\frac{1}{n}}z' + a(z^{\frac{1}{n}})z' + b(z^{\frac{1}{n}})z''z' = 0. \tag{1.17}$$

Therefore,

$$-c\frac{z^{1/n}}{b(z^{1/n})}z' + \frac{a(z^{1/n})}{b(z^{1/n})}z' + \frac{1}{2}\frac{d}{ds}(z')^2 = 0. \tag{1.18}$$

Consequently,

$$\frac{1}{2}\frac{d}{ds}(z')^2 - c\frac{d}{ds}A(z(s)) + \frac{d}{ds}B(z(s)) = 0, \tag{1.19}$$

where the functions $A(z)$ and $B(z)$ are defined by (1.3). A simple integration of (1.19) gives us:

$$\frac{(z')^2}{2} - cA(z) + B(z) = C_2. \tag{1.20}$$

Let us take $C_2 = 0$. The identity $z' = n\varphi^{n-1}\varphi'$ enables us to rewrite (1.20) in the form:

$$\frac{1}{2}n^2\varphi^{2(n-1)}(\varphi')^2 - cA(\varphi^n) + B(\varphi^n) = 0. \tag{1.21}$$

Evidently, $\varphi \equiv 0$ satisfies (1.21). Therefore, if $\varphi \not\equiv 0$ we have

$$\frac{n^2}{2}\varphi^{2n-2}\left[(\varphi')^2 - \frac{2c}{n^2}\frac{A(\varphi^n)}{\varphi^{2n-2}} + \frac{2}{n^2}\frac{B(\varphi^n)}{\varphi^{2n-2}}\right] = 0.$$

So we have reduced the problem of finding a travelling wave solution of the equation (1.2) to the solvability of the following ODE with separate variables:

$$(\varphi')^2 - 2A_1(\varphi) + 2B_1(\varphi) = 0. \tag{1.22}$$

The definitions of $A(z)$, $B(z)$ and (B) give us that near $z = 0$, $z \geq 0$: $A(z) \approx const\, z^{\frac{n+1}{n}}$ and $B(z) \approx const\, z^{\frac{m+n}{n}} \Rightarrow A(\varphi^n) \approx const\, \varphi^{n+1}$ and $B(\varphi^n) \approx \varphi^{n+m}$, $\varphi = z^{1/n}$, $z \approx 0$, $z \geq 0$ and the symbol const stands for some positive constants.

Another conclusion is that $A_1(\varphi) \approx const\, \varphi^{3-n}$, $\varphi \approx 0$, $const > 0$ and $B_1(\varphi) \approx const\, \varphi^{m-n+2}$, $const > 0$, $\varphi \approx 0$, $\varphi \geq 0$ (the functions $A_1(\varphi)$, $B_1(\varphi)$ are defined by (1.4)).

We can rewrite (1.22) as

$$(\varphi')^2 = 2(A_1(\varphi) - B_1(\varphi)) \tag{1.23}$$

and we know that according to (C):

$$A_1(\varphi) > B_1(\varphi), 0 < \varphi < D^{1/n}, A_1(D^{\frac{1}{n}}) = B_1(D^{\frac{1}{n}}), A_1'(D^{1/n}) \neq B_1'(D^{1/n}).$$

Certainly, we are loosing unicity for (1.23) as $\varphi \equiv D^{1/n}$ satisfies (1.23).
Consider now (1.23):

$$\varphi' = \sqrt{2[A_1(\varphi) - B_1(\varphi)]} \Rightarrow H(\varphi) = \int_0^{\varphi} \frac{d\lambda}{\sqrt{2(A_1(\lambda) - B_1(\lambda))}} = s,$$

i.e. $\varphi(0) = 0$, $H'(\varphi) \geq 0$.

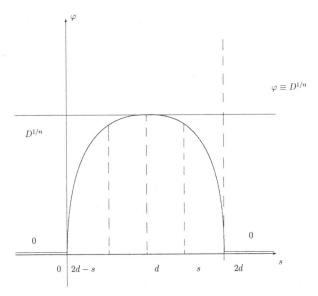

Fig. 1.3

The underintegral function has singularities at the points $\varphi = 0$, $\varphi = D^{1/n}$. Evidently, $H(\varphi) \approx const\, \varphi^{\frac{n-1}{2}}$, $\varphi \approx 0$, $\varphi > 0$ and $0 < H(D^{1/n}) = d < \infty$. In fact if we put $\Theta(z) = cA(z) - B(z)$, $0 < z < D$, then $\Theta(z) = (z - D)\Theta'(D) + o(|D - z|)$, $z \to D$ and according to (C): $\Theta'(D) < 0 \Rightarrow \Theta(\lambda^n) \approx const(D^{1/n} - \lambda)$ near $\lambda = D^{1/n}$, $const > 0$. This way we conclude that $H : [0, D^{1/n}] \to [0, d]$ is a homeomorphism, a diffeomorphism $(0, D^{1/n}) \to (0, d)$ and $H'(0) = 0$, $H'(D^{1/n}) = +\infty$.

Define now $\varphi = H^{-1}(s)$, $0 \le s \le d$. Then $\varphi(0) = 0$, $\varphi(d) = D^{1/n}$, $\varphi'(0) = +\infty$, $\varphi'(d) = 0$; $\varphi \approx const\, s^{\frac{2}{n-1}}$, $s \to +0$, $const > 0$.

To construct a compact travelling wave solution we shall continue $\varphi(s)$, $0 \le s \le d$ in the interval $[d, 2d]$ by the formula $\tilde{\varphi}(s) = H^{-1}(2d - s)$, $d \le s \le 2d$ (see Fig. 1.3) and we shall put $\varphi(\tilde{\varphi} = 0)$ for $s \le 0(s \ge 2d)$.

From geometrical point of view one can easily see that $V(s) = \begin{cases} \varphi(s), & s \in (0, d] \\ \tilde{\varphi}(s), & s \in [d, 2d) \end{cases}$, $V = 0$, $s \notin [0, 2d]$ is a smooth solution of (1.22), i.e. of (1.21) in $(0, 2d)$. Moreover, $V(s)$ satisfies (1.21) in a classical sense because near $s = 0(s = 2d)$: $\varphi^{n-1}\varphi' \approx const\, s^{2+\frac{3-n}{n-1}} = const\, s^{\frac{n+1}{n-1}}$, $s \to +0$. The travelling wave solution is given by: $u(t, x) = V(x - ct)$. The construction of $\varphi = H^{-1}(s)$ and of $\tilde{\varphi}$ is the same as the construction of a periodic solution of the mathematical pendulum. The new moment here is that $V'(+0) = \infty$, $V'(2d - 0) = -\infty$.

1.4 Generalizations of the results proved in Theorems 1.1 and 1.2

It is known from physics that the vibrations of a chain of particles inter-connected by springs is given by the following nonlinear PDE:

$$u_{tt} + \Phi'(u) = (T(u_x))_x, u = u(t, x). \tag{1.24}$$

This is the corresponding equation of the travelling wave solution $u = \varphi(x - ct)$:

$$c^2\varphi'' + \Phi'(\varphi) = [T(\varphi')]', \tag{1.25}$$

where T is a smooth function of the form $u_x + u_x^\alpha$, $\alpha > 1$ and $|u_x|$ is supposed to be small $(T(p) = p + p^\alpha, \alpha > 1)$.

Multiplying both sides of (1.25) by φ' we get

$$\frac{c^2}{2}\frac{d}{ds}(\varphi')^2 + \frac{d}{ds}\Phi(\varphi) = [T(\varphi')]'\varphi'. \tag{1.26}$$

Put $\tau(p) = \int T(p)dp$ and integrate (1.26) with respect to the variable s. We obtain

$$\frac{c^2}{2}(\varphi')^2 + \Phi(\varphi) = \varphi'T(\varphi') - \tau(\varphi') + C_1. \tag{1.27}$$

Taking $C_1 = 0$ we have

$$\frac{c^2}{2}(\varphi')^2 - \varphi'T(\varphi') + \tau(\varphi') = -\Phi(\varphi). \tag{1.28}$$

To simplify the things we shall suppose that the velocity c of the travelling wave solution is equal to 1, α is an odd integer, $\alpha \geq 3$ and $\Phi(\varphi) = \frac{1}{2}\varphi^2 - \frac{\varphi^{p+1}}{p+1}$, $p > 1$. Evidently, $\Phi(\varphi) \geq 0$ for $\varphi \geq 0$ iff $0 \leq \varphi \leq \varphi_2 = (\frac{p+1}{2})^{\frac{1}{p-1}} > 1$, $\Phi(\varphi) > 0$ for $0 < s < \varphi_2$.

Elementary computations show that in this case (1.28) takes the form:

$$(\varphi')^{\alpha+1} = \frac{\alpha+1}{\alpha}\Phi(\varphi), \tag{1.29}$$

i.e. $\varphi' = (\frac{\alpha+1}{\alpha})^{\frac{1}{\alpha+1}}[\Phi(\varphi)]^{\frac{1}{\alpha+1}}$ or $\varphi' = -(\frac{\alpha+1}{\alpha})^{\frac{1}{\alpha+1}}[\Phi(\varphi)]^{\frac{1}{\alpha+1}}$. Put $\frac{\alpha+1}{\alpha}\Phi(\varphi) = \Phi_1(\varphi)$. We are loosing unicity of the solutions of (1.29). Certainly, $\varphi \equiv \varphi_1 = 0$, $\varphi \equiv \varphi_2$ satisfies (1.29) and $0 \leq \varphi \leq \varphi_2$ in (1.29).

Consider now the nontrivial strictly monotonically increasing solution $H(\varphi) = s$ of (1.29), $\varphi(0) = 0$, where $0 < H(\varphi) = \int_0^\varphi \frac{d\lambda}{[\Phi_1(\lambda)]^{\frac{1}{\alpha+1}}}$, $0 < \varphi < \varphi_2$. The integral is convergent at the singular points $\varphi_1 = 0$, φ_2 and $H(0) = 0$, $0 < H(\varphi_2) = d < \infty$. Moreover, $\varphi \approx 0$, $\varphi > 0$ implies that $H(\varphi) \approx const\,\varphi^{\frac{\alpha-1}{\alpha+1}}$, $const > 0$, while $\varphi \approx \varphi_2$, $\varphi < \varphi_2$ implies that $H(\varphi) = d + \int_{\varphi_2}^{\varphi} \frac{d\lambda}{[\Phi_1(\lambda)]^{\frac{1}{\alpha+1}}} \approx d - const(\varphi_2 - \varphi)^{\frac{\alpha}{\alpha+1}}$, $const > 0$. In other words, $H : [0, \varphi_2] \to [0, d]$ is a homeomorphism, $H : (0, \varphi_2) \to (0, d)$ is a diffeomorphism and $H(\varphi) \approx const\,\varphi^{\frac{\alpha-1}{\alpha+1}}$ near $\varphi_1 = 0$, while $H(\varphi) \approx d - const(\varphi_2 - \varphi)^{\frac{\alpha}{\alpha+1}}$ near $\varphi = \varphi_2$. This way we construct $\varphi = H^{-1}(s)$, $0 \le s \le d$, $\varphi(0) = 0$, $\varphi'(0) = 0$, $\varphi'(d) = 0$ ($\varphi \approx s^{\frac{\alpha+1}{\alpha-1}}$ near $s = 0$ and $\varphi \approx \varphi_2 - const(d - s)^{\frac{\alpha+1}{\alpha}}$, $const > 0$, near $s = d$).

Our last step is to continue $\varphi(s)$ in the interaval $[d, 2d]$ as a smooth solution of (1.29): $\tilde{\varphi}(s) = \varphi(2d - s)$, $d \le s \le 2d$. This way we have obtained a smooth compact travelling wave solution of (1.29).

Remark 1.2. Under the assumption $|u_x|$ sufficiently small, i.e. $|\varphi'|$ small we can neglect $(\varphi')^{\alpha+1}$ in (1.28) and study (approximately) the equation: $\frac{1}{2}(c^2 - 1)(\varphi')^2 + \Phi(\varphi) = 0$, $|c| \ne 1$.

Remark 1.3. In the case $c = 1$, $\Phi(\varphi) = \varphi^\gamma$, $\gamma > \alpha + 1$, α — odd, we obtain in a standard way peakon but not compact travelling wave solution of (1.29) having $\varphi(0) = \bar{c}$, $\varphi(\infty) = 0$. In fact, it is easy to find that
$$\varphi(s) = (A|s| + \bar{c}^{1-\frac{\gamma}{\alpha+1}})^{\frac{-1}{\frac{\gamma}{\alpha+1}-1}}, \quad A = const > 0.$$

At the end of this section we shall say several words about the existence of compact travelling waves and peakon type solutions of the following generalized Camassa-Holm equation:

$$u_t - (u^n)_{xxt} + K(u^m)_x + \frac{1}{2}[g(u^p)]_x = \left[\frac{1}{2}((u^n)_x)^2 + u^n(u^n)_{xx}\right]'_x. \quad (1.30)$$

The case $n = m = p = 1$ was studied in [121]. To simplify the considerations we suppose that $n \ge 4$, $m > 1$, $0 < p \le n$; $m, n, p \in \mathbf{N}$, the function $g \in C^k(\mathbf{R}^1)$, with k — sufficiently large, $g(\lambda) > 0$ for $\lambda > 0$ and $g(\lambda) = \lambda^r(A + O(\lambda))$, $\lambda \to +0$, $A = const > 0$, $r \ge 1$, $pr \ge 2$.

The travelling wave solution $u = \varphi(x - ct)$, $c > 0$ of (1.30) will satisfy according to the investigations proposed in the proof Theorem 1.1 the ODE

$$c_2\varphi^{n+1} - K_2\varphi^{m+n} - \frac{1}{n^2}G(\varphi^p) = \varphi^{2n-2}(\varphi')^2(c - \varphi^n), \quad (1.31)$$

where $c_2 = \frac{2c}{n^2+n}$, $K_2 = \frac{2K}{n(m+n)}$, $G(z) = \frac{n}{p}\int_0^z g(\lambda)\lambda^{\frac{n}{p}-1}d\lambda$. Evidently, $G(z) > 0$ for $z > 0$, $G(z)$ is monotonically increasing for $z > 0$ and $G(z) = \frac{nA}{pr+n}z^{r+\frac{n}{p}}(1 + o(1))$, $z \to +0$.

As in (1.13) we shall construct a compacton-peakon type solution of (1.31) satisfying the ODE:

$$(\varphi')^2(c - \varphi^n) = \varphi^{3-n}\left[c_2 - K_2\varphi^{m-1} - \frac{G(\varphi^p)}{n^2\varphi^{n+1}}\right], \varphi(0) = \bar{c} = c^{1/n}. \quad (1.32)$$

Put $G_1(\varphi) = \frac{G(\varphi^p)}{n^2\varphi^{n+1}}$. Obviously, $0 < \varphi < c^{1/n} \Rightarrow G_1(\varphi) > 0$ and $G_1(\varphi) = \frac{A}{n(pr+n)}\varphi^{pr-1}(1 + O(\varphi))$, $\varphi \to +0$.

Denote $max_{0 \leq \varphi \leq c^{1/n}} G_1(\varphi) = B(c) > 0$ and assume that

$$c_2 - K_2 c^{\frac{m-1}{n}} - B(c) > 0, \quad (1.33)$$

i.e. $\frac{c}{n+1} - \frac{Kc^{\frac{m-1}{n}}}{m+n} - \frac{nB(c)}{2} > 0$.

Repeating the proof of Theorem 1.1 (see (1.15), (1.16)) we find a compacton-peakon solution of (1.30): $H_1(\varphi) = |s|$, where $H_1(\varphi) = \int_\varphi^{c^{1/n}} \frac{\sqrt{c-\lambda^n}\lambda^{\frac{n-3}{2}}d\lambda}{\sqrt{c_2-K_2\lambda^{m-1}-G_1(\lambda)}}$.

1.5 Cellular Neural Networks realization

At the begining of this paragraph we shall discuss briefly the Cellular Neural Network (CNN) approach. Cellular Neural Networks (CNNs) are complex nonlinear dynamical systems, and therefore one can expect interesting phenomena like bifurcations and chaos to occur in such nets. It was shown that as the cell self-feedback coefficients are changed to a critical value, a CNN with opposite-sign template may change from stable to unstable [114]. Namely speaking, this phenomenon arises as the loss of stability and the birth of a limit cycles [114].

CNN [25] is simply an analogue dynamic processor array, made of cells, which contain linear capacitors, linear resistors, linear and nonlinear controlled sources. Let us consider a two-dimensional grid with 3×3 neighbourhood system as it is shown on Fig. 1.4.

The squares are the circuit units — cells, and the links between the cells indicate that there are interactions between linked cells. One of the key features of a CNN is that the individual cells are nonlinear dynamical systems, but that the coupling between them is linear. Roughly speaking, one could say that these arrays are nonlinear but have a linear spatial structure, which makes the use of techniques for their investigation common in engineering or physics attractive.

We propose below the general definition of a CNN which follows the original one [25]:

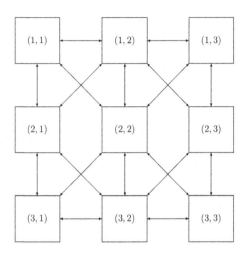

Fig. 1.4

Definition 1.2. The CNN is a

a) 2-, 3-, or n-dimensional array of

b) mainly identical dynamical systems, called cells, which satisfies two properties:

c) most interactions are local within a finite radius r, and

d) all state variables are continuous valued signals.

Definition 1.3. An $M \times M$ cellular neural network is defined mathematically by four specifications:

1) CNN cell dynamics;

2) CNN synaptic law which represents the interactions (spatial coupling) within the neighbour cells;

3) Boundary conditions;

4) Initial conditions.

Now in terms of Definition 1.2 we can present the dynamical systems describing CNNs. For a general CNN whose cells are made of time-invariant circuit elements, each cell $C(ij)$ is characterized by its CNN cell dynamics:

$$\dot{x}_{ij} = -g(x_{ij}, u_{ij}, I_{ij}^s), \qquad (1.34)$$

where $x_{ij} \in \mathbf{R}^m$, u_{ij} is usually a scalar. In most cases, the interactions (spatial coupling) with the neighbour cell $C(i + k, j + l)$ are specified by a

CNN synaptic law:

$$I_{ij}^s = A_{ij,kl}x_{i+k,j+l} + \tilde{A}_{ij,kl} * f_{kl}(x_{ij}, x_{i+k,j+l}) \tag{1.35}$$

$$+ \tilde{B}_{ij,kl} * u_{i+k,j+l}(t).$$

The first term $A_{ij,kl}x_{i+k,j+l}$ of (1.35) is simply a linear feedback of the states of the neighborhood nodes. The second term provides an arbitrary nonlinear coupling, and the third term accounts for the contributions from the external inputs of each neighbor cell that is located in the N_r neighborhood.

It is known [26] that some autonomous CNNs represent an excellent approximation to nonlinear partial differential equations (PDEs). The intrinsic space distributed topology makes the CNN able to produce real-time solutions of nonlinear PDEs. There are several ways to approximate the Laplacian operator in discrete space by a CNN synaptic law with an appropriate A-template. A one-dimensional discretized Laplacian template will be in the following form:

$$A_1 = (1, -2, 1).$$

This is the two-dimensional discretized Laplacian A_2 template:

$$A_2 = \begin{pmatrix} 0 & 1 & 0 \\ 1 & -4 & 1 \\ 0 & 1 & 0 \end{pmatrix}.$$

Definition 1.4. For any cloning template A which defines the dynamic rule of the cell circuit, we define the convolution operator $*$ by the formula

$$A * z_{ij} = \sum_{C(k,l) \in N_r(i,j)} A(k-i, l-j)z_{kl},$$

where $A(m,n)$ denotes the entry in the pth row and rth column of the cloning template, $p = -1, 0, 1$, and $r = -1, 0, 1$, respectively.

CNN model for the generalization of the Camassa-Holm equation (1.1) is written in the form:

$$\frac{du_{ij}}{dt} + KA_1 * u_{ij}^m - \frac{d}{dt}(A_2 * u_{ij}^n) = A_1 * \left[\frac{(A_1 * u_{ij}^n)^2}{2} + u_{ij}^n A_2 * u_{ij}^n \right], 1 \leq i, j \leq M. \tag{1.36}$$

For the generalized KdV equation (1.2) we obtain the following CNN model:

$$\frac{du_{ij}}{dt} + A_1 * a(u_{ij}) + A_1 * [b(u_{ij})A_2 * u_{ij}^m] = 0, 1 \le i,j \le M. \qquad (1.37)$$

There has been many studies on the travelling wave solutions of spatially discrete or both spatially and time discrete systems [78], [57]. In our case we shall study the existence and structure of the travelling wave solutions of autonomous Cellular Neural Networks. There are two possibilities of the structure of the travelling wave solutions — in one dimensional and two dimensional case respectively,

$$x_i = \Phi(i - ct),$$

or

$$x_{ij}(t) = \Phi(i\cos\Theta + j\sin\Theta - ct),$$

where $\Theta \in \mathbf{R}$ is given, $\Phi : \mathbf{R}^1 \to \mathbf{R}^1$ is a continuous function and c is unknown real number. Denote $s = i\cos\Theta + j\sin\Theta - ct$ (or $s = i - ct$). Then $\Phi(s)$ and c satisfy the equation of the form

$$-c\Phi'(s) = G(\Phi(s + r_0), \Phi(s + r_1), \ldots, \Phi(s + r_N)),$$

here $r_0 = 0$, r_i are real numbers for $i = 1$ to N. If the above equation depends on the past and future, i.e. if

$$r_{min} \equiv min\{r_i\}_{i=0}^N < 0 < r_{max} \equiv max\{r_i\}_{i=0}^N,$$

then it is called mixed type. If $r_{min} = 0$ or $r_{max} = 0$, then it is called advance or delay type respectively.

Our objective in this paper is to study the structure of the travelling wave solutions of the CNN models of generalized Camassa-Holm and KdV equations (1.36) and (1.37), respectively. We shall study the travelling wave solutions of the CNN models (1.36) and (1.37) of the form:

$$u_{ij} = \Phi(i\cos\Theta + j\sin\Theta - ct), \qquad (1.38)$$

for some continuous function $\Phi : \mathbf{R}^1 \to \mathbf{R}^1$ and for some unknown real number c. As we mentioned above $s = i\cos\Theta + j\sin\Theta - ct$. Let us substitute (1.38) in our CNN models (1.36) and (1.37). Therefore we consider solution $\Phi(s; c)$ of:

$$-c\Phi'(s; c) + G_1(\Phi(s; c)) = 0, \qquad (1.39)$$

$$-c\Phi'(s; c) + G_2(\Phi(s; c)) = 0, \qquad (1.40)$$

where $G_1(\Phi)$, $G_2(\Phi) \in \mathbf{R}^1$ are satisfying

$$lim_{s \to \pm\infty} \Phi(s;c) = 0. \tag{1.41}$$

for some $c > 0$. We shall investigate the basic properties of the solutions of (1.39) and (1.40).

Suppose that our CNN models (1.36) and (1.37) are finite circular arrays of $L = M.M$ cells. For this case we have finite set of frequences [104]:

$$\Omega = \frac{2\pi k}{L}, \ \ 0 \leq k \leq L - 1. \tag{1.42}$$

The following propositions then hold:

Proposition 1.1. *Suppose that $u_{ij}(t) = \Phi(i\cos\Theta + j\sin\Theta - ct)$ is a travelling wave solution of the CNN model (1.36) of the generalized Camassa-Holm equation (1.1) with $\Phi \in C^1(\mathbf{R}^1, \mathbf{R}^1)$ and $\Omega = \frac{2\pi k}{L}$, $0 \leq k \leq L-1$. Then there exist constants $c > 0$ and $s_0 > 0$ such that*

(i) for $s < s_0$ the solution $\Phi(s;c)$ of (1.39) satisfying (1.41) is increasing;

(ii) for $s > s_0$ the solution $\Phi(s;c)$ of (1.39) satisfying (1.41) is decreasing;

(iii) for $s = s_0$ the solution $\Phi(s;c)$ of (1.39) has maximum of cusp type or maximum of angle type with positive opening (peakon-cuspon, respectively peakon).

Moreover, the solution $\Phi(s;c)$ is either non vanishing everywhere or compactly supported, i.e. $\Phi(s;c) = 0$ for $|s-s_0| \geq d$, d being an appropriate positive constant.

Proposition 1.2. *Suppose that $u_{ij}(t) = \Phi(i\cos\Theta + j\sin\Theta - ct)$ is a travelling wave solution of the CNN model (1.37) of the generalized KdV equation (1.2) with $\Phi \in C^1(\mathbf{R}^1, \mathbf{R}^1)$ and $a(u) > 0$, $a(u) = u^m + O(u^{m+1}), u \to +0$, $b(u) > 0$ everywhere, $n \geq 4$, $m \geq n - 2$. Then there exist constants $c > 0$, $d > 0$ and finite set of frequences $\Omega = \frac{2\pi k}{L}$, $0 \leq k \leq L - 1$ such that $\Phi(s;c)$ of (1.40), (1.41) is vanishing outside the interval $[0, 2d]$, $\Phi(s;c)$ is increasing for $0 < s < d$ and $\Phi(s;c)$ is decreasing for $d < s < 2d$.*

Remark 1.4. There has been many studies on travelling wave solutions of spatially and time discrete systems [78], [57]. However, as far as we know travelling wave solutions of peakon type have been hardly studied in such discrete systems. For this reason we apply CNN approach and the numerical simulations of our CNN models (1.36) and (1.37) confirm the proposed results.

The simulations of the CNN models (1.36) and (1.37) give us the following Figs. 1.5, 1.6:

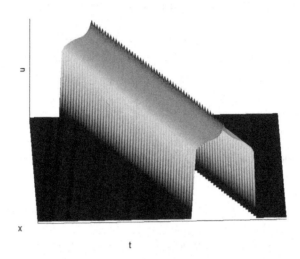

Fig. 1.5   The wave solution of (1.36) of the type compacton-peakon.

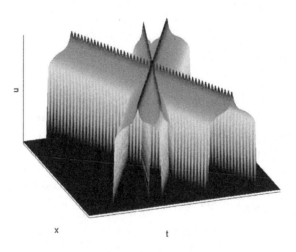

Fig. 1.6   Interaction of 3 travelling wave solutions of (1.37).

Chapter 2

On the existence and profiles of the travelling waves for several equations of mathematical physics

2.1 Introduction

Camassa-Holm considered (see [23], [24]) a third order nonlinear PDE of two variables modeling the propagation of unidirectional irrotational shallow water waves over a flat bad, as well as water waves moving over an underlying shear flow. As it is well known the motion of inviscid fluid with a constant density is described by the Euler's equations (system of nonlinear PDE). In the special case of the motion of a shallow water over a flat bottom the corresponding system was simplified by Green and Naghdi and related to an appropriate two component first order Camassa-Holm system. Another interesting system of non-linear PDE is the viscoelastic generalization of the Burger's equation. In the above mentioned systems we are looking for travelling wave solutions and we are studying their profiles. To do this we use several results from the classical Analysis of ODE that enable us to give the geometrical picture and in several cases to express the solutions by the inverse of the Legendre's elliptic functions. We are not going to discuss the physical explanation and interpretation and we shall concentrate only on the mathematical part of the investigations (see [96]).

2.2 Travelling waves for Camassa-Holm type equations

Camassa-Holm (see [23]) derived a shallow water equation

$$u_t + 2ku_x - u_{xxt} + 3uu_x = 2u_x u_{xx} + uu_{xxx} \qquad (2.1)$$

and proved that it possesses a peaked solitary wave solution for $k = 0$. Degasperis and Procesi (see [43], [55]) proposed the following variant of (2.1):

$$m_t + um_x + bu_x m = c_0 u_x - \gamma u_{xxx}, \qquad (2.2)$$

where $m = u - \alpha^2 u_{xx}$ and b, c_0, γ, α are constants, $\gamma \neq 0$, $\alpha \neq 0$. Thus, (2.2) can be written as

$$u_t - c_0 u_x + (b+1)uu_x - \alpha^2(u_{xxt} + uu_{xxx} + bu_x u_{xx}) + \gamma u_{xxx} = 0. \quad (2.3)$$

The travelling wave solution of (2.3) is:

$$u(x,t) = \Phi(x - ct) = \Phi(\xi), c = const., \xi = x - ct.$$

Substituting u in (2.3) we get:

$$-(c+c_0)\Phi' + (b+1)\Phi\Phi' - \alpha^2(\Phi\Phi''' + b\Phi'\Phi'') + (\alpha^2 c + \gamma)\Phi''' = 0. \quad (2.4)$$

Remark 2.1. Another possible generalization of (2.3) can be obtained by adding $u^r u_x = \Phi^r \Phi'$, $r \in \mathbf{N}$, $r \geq 2$ to the right hand side of (2.3). Then $\Phi^r \Phi' = \frac{1}{r+1}\frac{d}{d\xi}\Phi^{r+1}$; $\Phi'\Phi'' = \frac{1}{2}\frac{d}{d\xi}(\Phi')^2$, $\Phi\Phi''' = \frac{d}{d\xi}(\Phi\Phi'') - 1/2\frac{d}{d\xi}(\Phi')^2$.

Integrating (2.4) with respect to ξ we have:

$$-(c+c_0)\Phi + \frac{1}{2}(b+1)\Phi^2 - (\alpha^2\Phi - \alpha^2 c - \gamma)\Phi'' - \frac{\alpha^2}{2}(b-1)(\Phi')^2 + g = 0, \quad (2.5)$$

$g = const.$ Eventually, (2.5) contains the term $\frac{1}{r+1}\Phi^{r+1}$ (see Remark 2.1).

Our next step is to make in (2.5) the change $\Phi'_\xi = p(\Phi) \Rightarrow \Phi''_{\xi\xi} = \frac{1}{2}\frac{d}{d\Phi}(p^2)$. Put $p^2 = q \geq 0$. Then (2.5) takes the form

$$-(c+c_0)\Phi + \frac{1}{2}(b+1)\Phi^2 - \frac{1}{2}(\alpha^2\Phi - \alpha^2 c - \gamma)\frac{dq}{d\Phi} - \frac{\alpha^2}{2}(b-1)q + g = 0.$$

In more general form

$$(\alpha^2\Phi - \alpha^2 c - \gamma)\frac{dq}{d\Phi} + \alpha^2(b-1)q + 2(c+c_0)\Phi - (b+1)\Phi^2 + \varepsilon\Phi^{r+1} - 2g = 0, \quad (2.6)$$

where $\varepsilon^2 = \frac{2^2}{(r+1)^2}$ or $\varepsilon = 0$.

We shall concentrate on the case $\alpha^2\Phi - \alpha^2 c - \gamma \neq 0$, i.e. either $\Phi < \frac{\alpha^2 c + \gamma}{\alpha^2}$ or $\Phi > \frac{\alpha^2 c + \gamma}{\alpha^2}$.

The change of the independent variable Φ in (2.6): $\eta = \alpha^2\Phi - \alpha^2 c - \gamma$ leads to:

$$\eta\frac{dq}{d\eta} + (b-1)q + \frac{2(c+c_0)}{\alpha^4}(\eta + \alpha^2 c + \gamma) \qquad (2.7)$$

$$- (b+1)\frac{(\eta + \alpha^2 c + \gamma)^2}{\alpha^6} + \varepsilon\frac{(\eta + \alpha^2 c + \gamma)^{r+1}}{\alpha^{2(r+2)}} - \frac{2g}{\alpha^2} = 0.$$

There are no difficulties in studying the linear Euler type first order ODE (2.7). To do this we rewrite (2.7) as:

$$\eta\frac{dq}{d\eta} + (b-1)q + \sum_{\nu=0}^{r+1} a_\nu\eta^\nu = 0, \qquad (2.8)$$

a_ν being constants depending polynomially on $(\alpha^2 c + \gamma)$ and α^{-2} and on ε. If $\varepsilon = 0$ then we have a second order polynomial with respect to η in (2.8). Put $b - 1 = k \in \mathbf{R}$, $Q_{r+1}(\eta) = \sum_{\nu=0}^{r+1} a_\nu\eta^\nu \Rightarrow$

$$\eta\frac{dq}{d\eta} + kq + Q_{r+1}(\eta) = 0. \qquad (2.9)$$

The substitution $|\eta| = e^t$ in (2.9) enables us to conclude that $\frac{dq}{dt} + kq + Q_{r+1}(\pm e^t) = 0$, i.e.

$$q(t) = Ce^{-kt} + y_1(t), C = const, e^{-kt} = |\eta|^{-k}. \qquad (2.10)$$

Certainly, $\eta = e^t$ for $\eta > 0$, $\eta = -e^t$, $\eta < 0$, while $y_1(t)$ is some solution of

$$\frac{dy_1}{dt} + ky_1 + Q_{r+1}(\pm e^t) = 0.$$

Consequently,

$$y_1(t) = -e^{-kt}\int e^{kt}Q_{r+1}(\pm e^t)dt.$$

Therefore,

$$y_1(\eta) = -\sum_{\nu=0}^{r+1}\frac{a_\nu(\pm 1)^\nu e^{\nu t}}{\nu+k} = -\sum_{\nu=0}^{r+1}\frac{a_\nu(\pm 1)^\nu}{\nu+k}|\eta|^\nu \qquad (2.11)$$

for $k \neq 0, -1, \ldots, -(r+1)$, i.e. for $b \neq 1, 0, \ldots, -r$, while for $k = -\nu_0$, $\nu_0 = 0, 1, \ldots, (r+1)$; $(-k = \nu_0)$ we have

$$y_1(\eta) = -\sum_{\nu=0,\nu\neq -k}^{r+1}\frac{a_\nu(\pm 1)^\nu}{\nu+k}|\eta|^\nu - a_{\nu_0}(\pm 1)^{\nu_0}|\eta|^{\nu_0}ln|\eta|. \qquad (2.12)$$

Corollary 2.1. *Let $\varepsilon = 0$ in (2.7) and (2.8). Then these equations have the form:*

$$\eta\frac{dq}{d\eta} + kq + a_0 \pm a_1 e^t + a_2 e^{2t} = 0.$$

Thus $q = C|\eta|^{-k} + y_1$, where

$y_1 = -\frac{a_0}{k} + \frac{a_1}{k+1}|\eta| - \frac{a_2}{k+2}|\eta|^2$ *for* $k \neq \{0, -1, -2\}$,

$y_1 = -a_0 ln|\eta| - (\pm a_1)|\eta| - \frac{a_2}{2}|\eta|^2$, *for* $k = 0$

$y_1 = -(\pm a_1)|\eta|ln|\eta| - a_0 - a_2|\eta|^2$, *for* $k = -1$

$y_1 = +\frac{a_2}{2} - (\pm a_1)|\eta| - a_2|\eta|^2 ln|\eta|$, *for* $k = -2$.

Travelling wave solutions satisfy the ODE

$$\Phi'_\xi = p(\Phi) \Rightarrow (\Phi')^2 = p^2(\Phi) = q(\Phi),$$

i.e. we have an ODE with separate variables:

$$\frac{d\Phi}{d\xi} = \pm q(\Phi), q(\Phi) \geq 0, \frac{d\Phi}{\pm q(\Phi)} = d\xi \Rightarrow \pm \int \frac{d\Phi}{q(\Phi)} = \xi - \xi_0$$

and according to (2.10) $q(\eta) = C|\eta|^{-k} + y_1(\eta)$ (see (2.11), (2.12)), $\eta = \alpha^2 \Phi - \alpha^2 C - \gamma$. A detailed study of the solutions of the ODE $(\Phi')^2 = q(\Phi)$ is given in Chapter III.

The travelling wave solution $\Phi(\xi)$ can be expressed by elliptic functions if $\varepsilon = 0$ in (2.7) and say $k = -4, -3$, as then $q(\eta)$ is a polynomial with respect to Φ of order 4 or 3. ($k = -3, -4 \Rightarrow b - 1 = k = -3, -4 \Rightarrow b = -2, -3$.) If $k = -2 \Rightarrow b = -1$ we must take $a_2 = 0$ in $y_1(\eta)$, while for $k = -1 \Rightarrow b = 0$ we must take $a_1 = 0$ in $y_1(\eta)$. In both cases appears the integral $\int \frac{d\Phi}{\sqrt{A+B\Phi+C\Phi^2}}$ with appropriate constants A, B, C. Certainly, we shall have elliptic integrals if $\varepsilon \neq 0$, $k = -3, -4$; $r = 2, 3$ but with the assumption $k = -3 \Rightarrow a_{-k} = 0$; $k = -4 \Rightarrow a_{+4} = 0$.

2.3 Travelling waves for viscoelastic generalization of the Burger's equation

In the paper [22] travelling waves and shocks in viscoelastic generalization of Burger's equation are studied. The wave profiles are given by using the numerical approach and by giving the corresponding computer visualizations. More precisely, in [22] shock waves could appear too. We propose in this paragraph a purely mathematical approach to the same problem and give a qualitative picture of the behavior of these waves. To do this we use elementary facts from the theory of ODE.

In [22] the following generalization of the Burger's equation is investigated:

$$\begin{cases} u_t + uu_x = \sigma_x, \alpha, \beta = const > 0 \\ \sigma_t + u\sigma_x - \sigma u_x = \alpha u_x - \beta\sigma. \end{cases} \tag{2.13}$$

The function σ stands for the stress, u stands for velocity and (2.13) describes how the addition of viscoelasticity affects travelling wave solutions of Burger's equation. If there is no relaxation of stress then one takes $\beta = 0$.

Put now $u(x,t) = U(\xi)$, $\xi = x - ct$, $c = const$, $\sigma(x,t) = S(\xi)$ and substitute them in (2.13). Thus,

$$\begin{cases} -cU' + UU' = S' \\ -cS' + US' - SU' = \alpha U' - \beta S. \end{cases} \tag{2.14}$$

Integrating the first equation of (2.14) we get

$$S = \frac{1}{2}U^2 - cU + A, A = const$$

and therefore the second equation implies

$$c^2 U' - cUU' + U(U'U - cU') - \left(\frac{1}{2}U^2 - cU + A\right)U'$$

$$= \alpha U' - \beta\left(\frac{1}{2}U^2 - cU + A\right),$$

i.e.

$$U'\left[U\left(\frac{U}{2} - c\right) + c^2 - A - \alpha\right] = -\beta\left[U\left(\frac{U}{2} - c\right) + A\right]. \tag{2.15}$$

Remark 2.2. In [22] the authors assume that $U(-\infty) = u_l$, $S(-\infty) = 0$; $U(\infty) = u_r$, $S(\infty) = 0$, $u_r \neq u_l$. Then $S(\pm\infty) = 0$ implies that $0 = \frac{1}{2}u_l^2 - cu_l + A$, $0 = \frac{1}{2}u_r^2 - cu_r + A \Rightarrow c = \frac{u_r + u_l}{2}$, $A = \frac{1}{2}u_l u_r$ and (2.15) takes the form

$$U'\left[(U-u_r)(U-u_l)+2\left(\left(\frac{u_r - u_l}{2}\right)^2 - \alpha\right)\right] = -\beta(U-u_r)(U-u_l). \tag{2.16}$$

To simplify the things we denote $k = 2(\frac{(u_r-u_l)^2}{4} - \alpha)$ and consider several different cases having in mind that if $u_r < u_l \Rightarrow u_r < U < u_l \Rightarrow f(U) = (U - u_r)(U - u_l) < 0$, $f(u_r) = f(u_l) = 0$, $f(U) \in [-\frac{(u_l-u_r)^2}{4}, 0]$.

Moreover, $min_{U \in [u_r,u_l]} f(U) = f(\frac{u_r+u_l}{2}) = -\frac{(u_r-u_l)^2}{4}$. Thus,

1) $k < 0 \Rightarrow f(U) + k < 0$, $\forall U \in [u_r, u_l]$ (i.e. $\alpha > \frac{(u_r-u_l)^2}{4}$).

2) $k > 0$. Then $f(U) + k \in [-\frac{(u_l-u_r)^2}{4} + k, k]$, $\forall U \in [u_r, u_l]$ and we take $k > \frac{(u_l-u_r)^2}{4} \iff \alpha < \frac{1}{2}(\frac{u_r-u_l}{2})^2$.

3) $0 < k < \frac{(u_l-u_r)^2}{4} \iff \frac{(u_l-u_r)^2}{8} < \alpha < \frac{(u_l-u_r)^2}{4}$.

4) $k = 0 \Rightarrow (U' + \beta)(U - u_r)(U - u_l) = 0$; $k = \frac{(u_l-u_r)^2}{4}$.

In cases 1), 2) we have classical monotone solutions of (2.16), while in 3) the picture is rather interesting as then shock waves of the system (2.13) equipped with appropriate initial data can appear [22]. The corresponding

solutions U of (16) are multivalued and therefore, no classical travelling
wave solutions exist.

Consider the case 3). Geometrically, the straight line $Z = -k$ is crossing
twice the parabola $Z = f(U) = (U - u_r)(U - u_l)$ at the points $(u_{1,2}, -k)$,
$u_1 \neq u_2$.

Certainly, we have: $f(u_1) + k = 0$, $f(u_2) + k = 0$, $u_r < u_1 < \frac{u_r + u_l}{2} < u_2 < u_l$,

$f(U) + k > 0$ for $U \in [u_r, u_1)$,

$f(U) + k > 0$ for $U \in (u_2, u_l]$ and

$U \in (u_1, u_2) \Rightarrow f(U) + k < 0$.

Moreover, if $k = \frac{(u_l - u_r)^2}{4}$ then $Z = -k$ is tangential to the parabola
$Z = f(U)$ at its vertex. Then $\alpha = \frac{(u_r - u_l)^2}{8}$, $u_1 = u_2 = \frac{u_r + u_l}{2}$.

In the case 1) $U' < 0$ for $U \in (u_r, u_l)$ and therefore U is strictly
monotonically decreasing function such that $U(-\infty) = u_l$, $U(+\infty) = u_r$
$\Rightarrow u_l > u_r$. This is so called kink solution.

On the other hand in case 2) $U' > 0$ and therefore U is strictly mono-
tonically increasing for each ξ, i.e. $U(-\infty) = u_l$, $U(\infty) = u_r \Rightarrow u_l < u_r$.
$U = u_r$, $U = u_l$ are horizontal asymptotes of the solution. Certainly, such
situation is impossible in our case.

In case 3) there exists a multivalued (three valued) continuous solution
of (2.16) which is smooth up to two points of its graph where it possesses
vertical tangents. This solution will be constructed in three steps. In the
first step let $u_r < U_0 < u_1$ and $\xi_0 \in \mathbf{R}$. Then for $U \in (u_r, u_1) \Rightarrow U' > 0$.
The right hand side of (2.16) is C^1 for $U \in [u_r, u_1)$ and consequently

$$\xi - \xi_0 = -\int_{U_0}^{U} \frac{k + (\lambda - u_r)(\lambda - u_l)d\lambda}{(\lambda - u_r)(\lambda - u_l)\beta} \equiv F(U) \in C^1(u_r, u_1]. \qquad (2.17)$$

Evidently, $F'(U) > 0$ for $U \in [u_r, u_1)$, $F(U_0) = 0$, $lim_{U \to u_r}F(U) = -\infty$
as the integral is divergent at $U = u_r$, $lim_{U \to u_1}F(U) = F(u_1) > 0$,
$F'(u_1) = 0$. Thus, $F : (u_r, u_1) \to (-\infty, F(u_1))$ is diffeormorphism, and
homeomorphism $F : (u_r, u_1] \to (-\infty, F(u_1)]$, $\xi - \xi_0 = F(U) \Rightarrow \xi = \xi_0 + F(U) \in (-\infty, F(u_1) + \xi_0)$; if we put $\bar{\xi} = F(u_1) + \xi_0 > \xi_0$ then
$U = F^{-1}(\xi - \xi_0)$, $\xi \in (-\infty, \bar{\xi})$, $U(\xi_0) = F^{-1}(0) = U_0$, $lim_{\xi \to -\infty}U(\xi) = u_r$,
$lim_{\xi \to \bar{\xi}}U(\xi) = U(\bar{\xi}) = u_1$, $U'(\bar{\xi}) = +\infty$.

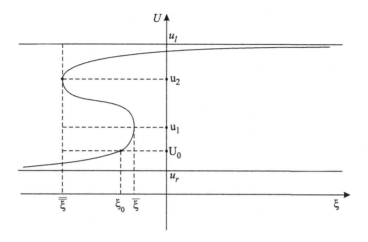

Fig. 2.1

In the second step we construct a solution with $u_1 < U < u_2 \Rightarrow U' < 0$ and initial data $(\bar{\xi}, u_1)$. Then (2.17) takes the form

$$\xi - \bar{\xi} = -\int_{u_1}^{U} \frac{k + (\lambda - u_r)(\lambda - u_l)d\lambda}{(\lambda - u_r)(\lambda - u_l)\beta} = F_1(U) \in C^1[u_1, u_2] \qquad (2.18)$$

as the underintegral function has not singularities. Thus, $\xi - \bar{\xi} = F_1(U) < 0$, $F_1(u_1) = 0$, $U(\bar{\xi}) = u_1$, $F_1'(u_1) = 0$, $F_1'(u_2) = 0$, i.e. if we put $\bar{\bar{\xi}} = \bar{\xi} + F_1(u_2)$ the mapping

$F_1 : [u_1, u_2] \to [F_1(u_2), F_1(u_1)]$ is a homeomorphism,

while $F_1 : (u_1, u_2) \to (F_1(u_2), F_1(u_1))$ is diffeomorphism

and $U'(\bar{\xi}) = -\infty$, $U'(\bar{\bar{\xi}}) = -\infty$, $\bar{\bar{\xi}} < \bar{\xi}$, $U(\bar{\bar{\xi}}) = u_2$, $U = F_1^{-1}(\xi - \bar{\xi})$.

The third step is standard as the we construct a solution U, $U' > 0$ passing through $(\bar{\bar{\xi}}, u_2)$. Evidently, $U'(\bar{\bar{\xi}}) = +\infty$, $U(\xi) \in C^1(\bar{\bar{\xi}}, +\infty)$, $lim_{\xi \to \infty} U(\xi) = u_l$ (see Fig. 2.1).

It is interesting to point out that the conditions at $\pm\infty$ are not satisfied by U as $U(-\infty) = u_r$, $U(\infty) = u_l$. Moreover, U is triple valued for $\xi \in (\bar{\xi}, \bar{\bar{\xi}})$.

Case 4. Let now $k = \frac{(u_l - u_r)^2}{4} \Rightarrow U' > 0$. The solution is again single valued and has a vertical tangent at only one point ($u_1 = u_2 = \frac{u_r + u_l}{2}$, $\bar{\xi} = \bar{\bar{\xi}}$) (then $\alpha = \frac{(u_r - u_l)^2}{8}$).

Again $U(-\infty) = u_r$, $U(+\infty) = u_l$. If $k = 0$, i.e. $\alpha = \frac{(u_r - u_l)^2}{4}$ the solution U is piecewise linear function (see Fig. 2.2).

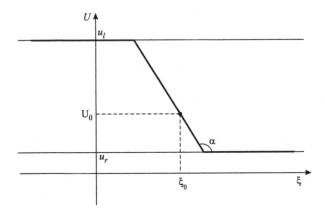

Fig. 2.2

2.4   Travelling wave solutions for the two component Camassa-Holm type system

The motion of inviscid fluid with a constant density is described by the well-known Euler's equations. On the other hand, the motion of a shallow water over a flat bottom is described by a 3×3 semilinear system of first order partial differential equations. Because of the lack of space we do not give the exact expressions of the above mentioned systems. Starting from the semilinear 3×3 system one can obtain the so-called Green-Naghdi system considered for the first time in 1976. The latter system can be related to the following two component Camassa-Holm type system:

$$
\left|
\begin{aligned}
&m_t + 2u_x m + u m_x + \rho \rho_x = 0 \\
&\rho_t + (u\rho)_x = 0, \text{where} \quad m = u - u_{xx}.
\end{aligned}
\right.
\tag{2.19}
$$

We are looking for travelling wave solutions of (2.19), i.e. $u(t, x) = u(\xi)$, $m(t, x) = m(\xi)$, $\xi = x - ct$, $c = const$. Substituting $u(\xi)$, $m(\xi)$ in (2.19) we get from the second equation that $-c\rho' + (u\rho)' = 0 \Rightarrow c\rho(\xi) = u(\xi)\rho(\xi) - \alpha$, $\alpha = const \Rightarrow \rho = \frac{\alpha}{u-c}$; $m = u - u''$. Therefore, the first equation in (2.19) implies:

$$
-cm' + (um)' + u'm + \rho\rho' = 0
$$

$$
\Rightarrow -c(u' - u''') + (um)' + u'(u - u'') + \rho\rho'' = 0
$$

and after an integration with respect to ξ we get that

$$-cu + cu'' + u^2 - uu'' + \frac{1}{2}u^2 - \frac{1}{2}(u')^2 + \frac{1}{2}\rho^2 = \frac{\beta}{2} = const,$$

i.e.

$$-cu + cu'' + \frac{3}{2}u^2 - uu'' - \frac{1}{2}(u')^2 + \frac{\alpha^2}{(u-c)^2} = \frac{\beta}{2}. \qquad (2.20)$$

Assuming $c > 0$ we make in (2.20) the change $u = c(1 + z)$. In the case $c < 0$ we shall make the change $u = c(1 - z)$. We shall confine ourselves to the case when $c > 0$ and $z = \frac{u}{c} - 1$ as the other case is treated similarly. Consequently, $\rho = \frac{\alpha}{cz}$, $z \neq 0$ and then (2.20) can be written as:

$$-c^2(1 + z) + c^2 z'' + \frac{3}{2}c^2(1+z)^2 - c^2(1+z)z'' - \frac{1}{2}c^2(z')^2 + \frac{1}{2}\frac{\alpha^2}{c^2 z^2} = \frac{\beta}{2}$$

$$\Rightarrow zz'' + \frac{1}{2}(z')^2 - \frac{3}{2}z^2 - 2z - \frac{\alpha^2}{2c^4 z^2} + \frac{\beta - c^2}{2c^2} = 0. \qquad (2.21)$$

We multiply (2.21) by z' then integrate in ξ and having in mind that $zz'z'' + \frac{1}{2}(z')^3 = \frac{1}{2}(z(z')^2)'$ we obtain:

$$\frac{1}{2}z(z')^2 - \frac{z^3}{2} - z^2 + \frac{\alpha^2}{2c^4 z} + \frac{\beta - c^2}{2c^2}z = \frac{\gamma}{2} = const,$$

i.e.

$$z^2(z')^2 = z^4 + 2z^3 + \frac{c^2 - \beta}{c^2}z^2 + \gamma z - \frac{\alpha^2}{c^4}. \qquad (2.22)$$

Put $\frac{c^2-\beta}{c^2} = \beta_1 \Leftrightarrow \beta = c^2(1 - \beta_1)$, $-\frac{\alpha^2}{c^4} = \alpha_1 < 0 \Leftrightarrow \alpha = \pm c^2\sqrt{-\alpha_1}$, i.e. $\alpha \to 0 \Leftrightarrow \alpha_1 \to 0$.

Let us fix $c > 0$. Then $\beta_1 \in \mathbf{R}$, $\gamma \in \mathbf{R}$ and $\alpha_1 < 0$ are arbitrary real constants.

Put $P_4(z) = z^4 + 2z^3 + \beta_1 z^2 + \gamma z + \alpha_1$.

If $\gamma = 0$ we write $\tilde{P}_4(z) = z^4 + 2z^3 + \beta_1 z^2 + \alpha_1$. Define now $w = \tilde{\tilde{P}}_4(z) = z^2(z^2 + 2z + \beta_1)$. Evidently, $\tilde{\tilde{P}}_4(z) = 0 \Leftrightarrow z_{1,2} = 0$, $z_{3,4} = -1 \pm \sqrt{1 - \beta_1}$. We shall assume further on that

$$0 < \beta_1 < 1 \Leftrightarrow 0 < 1 - \beta_1 < 1 \Leftrightarrow -1 < z_3 < 0, -2 < z_4 < -1. \qquad (2.23)$$

Consider now the cross points of the biquadratic parabola $w = \tilde{\tilde{P}}_4(z)$ and the straight line $w = -\alpha_1 > 0$, $0 < -\alpha_1 \ll 1 \Leftrightarrow 0 < |\alpha| \ll 1$. Geometrically we have Fig. 2.3.

Evidently, the curve $w = \tilde{\tilde{P}}_4(z)$ crosses the line $w = -\alpha_1$, $|\alpha_1| \ll 1$ at the points $z_4' < z_4 < z_3 < z_3' < z_2' < 0 < z_1'$. Therefore $\tilde{P}_4(z) = 0 \Leftrightarrow \tilde{P}_4(z_j') = 0$,

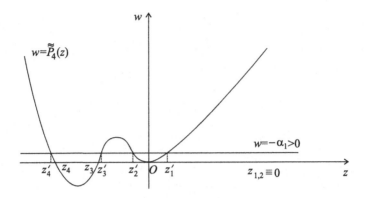

Fig. 2.3

$z_{j,\,1 \le j \le 4}$ being 4 simple roots of the algebraic equation $\tilde{P}_4(z) = 0$. As it is known from Analysis, the simple roots of the algebraic equations depend continuously on the coefficients of the corresponding polynomials. This way we come to the Proposition 2.1.

Proposition 2.1. *Consider the 4th order polynomial $P_4(z)$, fix the constant $c > 0$ and suppose that $0 < \beta_1 < 1$. Then one can find some $0 < \varepsilon_0$ such that if $|\gamma| \le \varepsilon_0$, $|\alpha| \le \varepsilon_0$, then $P_4(z) = 0$ has 4 simple roots $z_4' < z_3' < z_2' < 0 < z_1'$.*

According to [10], [90], [50] the equation

$$\left| \begin{array}{l} (z')^2 = \dfrac{P_4(z)}{z^2} \ge 0 \\ z(0) = z_0 \in [z_3', z_2'] \end{array} \right. \tag{2.24}$$

possesses a smooth periodic solution $z(\xi)$, $z(0) = z_0$, $z_3' \le z(\xi) \le z_2'$, i.e. $z \ne 0$. In fact, $P_4(z) \ge 0$ for $z \in [z_3', z_2']$ (see Fig. 2.3), while $P_4(z) < 0$ for $z \in (z_4', z_3')$ or $z \in (z_2', z_1')$. At first we construct the solution of $z' = \dfrac{\sqrt{P_4(z)}}{-z} > 0$, $z(0) = z_0$ as $z \in [z_3', z_2'] \Rightarrow -z > 0$, i.e.

$$\xi = \int_{z_3'}^{z} \frac{-\lambda d\lambda}{\sqrt{P_4(\lambda)}} = H(z) \quad \text{and} \quad \text{for } z_0 = z_3'. \tag{2.25}$$

Evidently, $H'(z) > 0$ for $z \in (z_3', z_2')$, $H'(z_3') = \infty$, $H'(z_2') = \infty$, i.e. $\left(H^{-1}(\xi)\right)' > 0$, $H(z_3') = 0$. Put $0 < \dfrac{T}{2} = \int_{z_3'}^{z_2'} \dfrac{-\lambda d\lambda}{\sqrt{P_4(\lambda)}} \Rightarrow H(z_2') = \dfrac{T}{2}$, $z = z(\xi)$, $0 \le \xi \le \dfrac{T}{2}$, $z'(0) = z'\left(\dfrac{T}{2}\right) = 0$. Then we continue $z(\xi)$ as an even function on the interval $-\dfrac{T}{2} \le \xi \le 0$, i.e. $z(-\xi) = z(\xi)$. One can see easily that $z(\xi)$ satisfies the ODE $z'^2 = \dfrac{P_4(z)}{z^2}$ on $\left[-\dfrac{T}{2}, \dfrac{T}{2}\right]$. Our last step is to continue $z(\xi)$ as a smooth periodic function with period T on the real line $\mathbf{R}_\xi^1$. According to [100], p. 257, for $z_3' \leq z \leq z_2'$

$$-\xi = -H(z) = \int_{z_3'}^{z} \frac{\lambda d\lambda}{\sqrt{P_4(\lambda)}} = \frac{2}{\sqrt{(z_1' - z_3')(z_2' - z_4')}}$$

$$\times \left\{ (z_3' - z_4') \, \Pi\left(\delta, \frac{z_2' - z_3'}{z_2' - z_4'}, q \right) + z_4' F(\delta, q) \right\},$$

where $q = \sqrt{\frac{(z_2' - z_3')(z_1' - z_4')}{(z_1' - z_3')(z_2' - z_4')}}$, $\delta = \arcsin\sqrt{\frac{(z_2' - z_4')(z - z_3')}{(z_2' - z_3')(z - z_4')}}$ and F, Π are Legendre's elliptic integrals of first and third kind respectively (see [21], [100]). Certainly, $H(z) = |\xi|$ for $|\xi| \leq \frac{T}{2}$, $z_3' \leq z \leq z_2'$.

This way we expressed a class of periodic solutions of (2.19) – travelling wave type – by the famous Legendre's elliptic functions. In fact $z \neq 0$ is periodic with period T and $u = c(1+z)$, $\rho = \frac{\alpha}{cz}$.

In Chapter III we are going to propose a short excursion in the theory of elliptic functions including F and Π elliptic integrals, the elliptic sn and cn functions, Weierstrass functions etc. The previous example can be considered as a simple application of the Legendre's elliptic functions to the travelling wave solutions for the two component Camassa-Holm system. The reader can find several illustrative examples and many comments in Chapter III.

Remark 2.3. Consider the ODE (2.24), where $P_4(z) = z^4 + 2z^3 + \beta_1 z^2 + \gamma z + \alpha_1$, β_1, γ, $\alpha_1 < 0$ being arbitrary constants. Certainly, $P_4(z) = 0$ possesses at least two real roots as $P_4(0) = \alpha_1 < 0$. We assume that $k_1 = k_2 < 0$ is a double root of $P_4(z)$, while $k_1 < k_3 < 0 < k_4$ are simple roots. We have supposed that $k_3 < 0 < k_4$ as $k_1^2 k_3 k_4 = \alpha_1 < 0$. Therefore, $k_1 < z < k_3 \Rightarrow P_4(z) > 0$. Then the Cauchy problem

$$\begin{vmatrix} z^2 z'^2 = P_4(z) = (z - k_1)^2 (z - k_3)(z - k_4) \\ z(0) = k_3 \end{vmatrix} \tag{2.26}$$

possesses a smooth solution $z(\xi)$, $\xi \in \mathbf{R}^1$, such that $z(-\xi) = z(\xi)$, $\forall \xi \in \mathbf{R}^1$, $z'(0) = 0$, $z'(\xi) < 0$ for $\xi > 0$. This is a soliton, of course (see Fig. 2.4). Moreover, we can give an explicit formula for the solution as the corresponding integral $\xi = \int_{k_3}^{z} \frac{\lambda d\lambda}{(\lambda - k_1)\sqrt{(\lambda - k_3)(\lambda - k_4)}} \equiv H_1(z) > 0$, $k_1 < z < k_3$, $H_1'(z) < 0$, $H_1(k_3) = 0$, $\lim_{z \to k_1} H_1(z) = +\infty$, $H_1'(k_3) = -\infty$ can be calculated by using the standard Euler's substitutions.

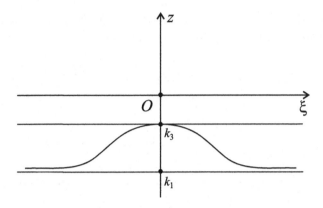

Fig. 2.4

The case when $P_4(z)$ has a triple root or a pair of complex roots w, $\overline{w} \in \mathbf{C}^1 \backslash \mathbf{R}^1$ can be studied in a similar way and we omit the details. They are left to the reader. We complete our study. We are ready now to discuss more general problems concerning larger classes of travelling wave solutions.

Chapter 3

Travelling wave solutions of special type to third order nonlinear PDE of mathematical physics

3.1 Introduction and formulation of the main results

As it is well known and it was mentioned in Chapters I, II the Camassa-Holm equation

$$u_t - u_{xxt} + 3uu_x = 2u_x u_{xx} + uu_{xxx}, t > 0, x \in \mathbf{R} \tag{3.1}$$

models the propagation of water waves in the shallow water regime when the wave length is considerably larger than the average water depth ([23], [29], [49], [19]). Recently the following Camassa-Holm — γ (CH-γ) equation was studied by many authors ([44], [75], [112]):

$$u_t + c_0 u_x + 3uu_x - \alpha^2(u_{xxt} + uu_{xxx} + 2u_x u_{xx}) + \gamma u_{xxx} = 0, \tag{3.2}$$

where α^2, c_0 and γ are real-valued parameters.

As we mentioned in Chapter I, Rosenau has considered several generalizations of the Korteweg-de Vries (KdV) equation ([101], [102], [103]) and has shown that they possess a new form of solitary wave with compact support. This is for example the equation

$$u_t + (u^n)_x + (u^n)_{xx} = 0, \tag{3.3}$$

where n is a parameter. More precisely, the function $U(x - ct)$, where $U(\xi) = [\frac{cn}{n+1} cos^2(\frac{n-1}{2n}\xi)]^{\frac{1}{n-1}}$ for $|\xi| \le \frac{n}{n-1}$, $c > 0$, $U = 0$ elsewhere satisfies the PDE (3.3), respectively, $U(\xi)$ satisfies the corresponding ODE on the whole real line.

A different generalization of the KdV equation was studied in [35], [62], [108], namely:

$$u_t = u_x u^{l-2} + \alpha[2u_{xxx}u^p + 4pu^{p-1}u_x u_{xx} + p(p-1)u^{p-1}(u_x)^3], \qquad (3.4)$$

where p, l, α are constants.

In Chapter I ([97]) we investigated the following generalization of Camassa-Holm equation:

$$u_t + K(u^m)_x - (u^n)_{xxt} = \left[\frac{((u^n)_x)^2}{2} + u^n(u^n)_{xx}\right]_x, \qquad (3.5)$$

where $K = const > 0$ and $m, n \in \mathbf{N}$.

As we show in Chapter I smooth compact travelling wave solutions exist for the equation describing the vibrations of a chain of particles interconnected by springs:

$$u_{tt} + \Phi'(u) = (T(u_x))_x, \qquad (3.6)$$

where $T(p) = p + p^\alpha$, $\alpha > 1$.

Recently the cusped solutions of the CH equation, the cuspon solitary waves and the antipeakon limits were considered in [89]. Therefore, there are many papers devoted to the existence of travelling wave solutions of third order nonlinear PDE of mathematical physics. They can be smooth or they can develop singularities of the following two types: peaks, cusps. We remind to the reader that in the case of peak singularity at a point the solution has left and right hand sides tangents there forming an angle with positive opening.

To be more precise, the travelling wave solutions $u = \varphi(x - ct)$, $c = const > 0$ of the above mentioned equations satisfy ODE with separate variables and of special type. We propose below several of them.

In the case of CH- γ equation (3.2) the corresponding ODE is:

$$(\varphi')^2(\gamma + c\alpha^2 - \alpha^2\varphi) = -\varphi^3 - (c_0 - c)\varphi^2 - C_1\varphi - C_2, \qquad (3.7)$$

where C_1, C_2 are arbitrary constants; $\varphi = \varphi(\xi)$, $\xi = x - ct$. As it concerns (3.4), then we have the ODE:

$$\alpha\varphi^p(\varphi')^2 + \frac{\varphi^l}{l(l-1)} - \frac{c}{2}\varphi^2 + C_1\varphi + C_2 = 0, \qquad (3.8)$$

where $l \neq 0, 1$ and C_1, C_2 are arbitrary constants.

The generalized CH equation (3.5) is reduced to the following ODE:

$$(\varphi')^2(c - \varphi^n) = c_1\varphi^{3-n} - K_1\varphi^{m+2-n} + C_1\varphi^{2-n} + C_2\varphi^{2-2n}, \qquad (3.9)$$

where $c_1 = \frac{2c}{n^2+n}$, $K_1 = \frac{2K}{m(m+n)}$ and again C_1, C_2 are arbitrary constants.

All these equations can be generalized in a simple way. Thus, consider the nonlinear third order PDE of 2 independent variables (t, x):

$$F(u, u_t, u_x, u_{tt}, u_{tx}, \ldots, u_{xxx}) = 0, u = u(t, x). \qquad (3.10)$$

Putting $u = \varphi(x - ct)$, $c \neq 0$, $\varphi = \varphi(\xi)$, $\xi = x - ct$ we obtain:

$$A(\varphi) = F(\varphi, -c\varphi', \varphi', c^2\varphi'', -c\varphi'', \ldots, \varphi''') = 0.$$

We shall assume further on that for some $c \neq 0$

$$A(\varphi) = \frac{d}{d\xi}G_c(\varphi, \varphi', \varphi''), \qquad (3.11)$$

where G_c is a smooth function depending on the constant c.

Thus, $G_c(\varphi, \varphi', \varphi'') = C_1 = const.$ Multiplying G_c by φ' we get:

$$\varphi' G_c(\varphi, \varphi', \varphi'') = C_1\varphi'. \qquad (3.12)$$

This is our main assumption:

$$\varphi' G_c(\varphi, \varphi', \varphi'') = \frac{d}{d\xi}(P_1(\varphi)(\varphi')^2 + P_2(\varphi)), \qquad (3.13)$$

where $P_1(\varphi)$, $P_1(\varphi)$ are polynomials of φ.

So we obtain that

$$P_1(\varphi)(\varphi')^2 + P_2(\varphi) = C_1\varphi + C_2, \qquad (3.14)$$

C_1, C_2 being arbitrary constants.

A glance at (3.7)–(3.9) shows that in the case l, p, m, n-integers these equations are of the form (3.14) for appropriate P_1 and P_2. Therefore, (3.14) is a possible generalization of several ODE arising in finding of the travelling wave solutions of some equations of mathematical physics (see [95]).

2. This way we come to the necessity of studying of the following Cauchy problem:

$$\begin{cases} Q(y)(y')^2 = P(y), y = y(x) \\ y(x_0) = y_0, \end{cases} \qquad (3.15)$$

where $P(y)$, $Q(y)$ are polynomials of y, $Q(y)P(y) > 0$, $\forall y \in (\alpha, \beta)$ and without loss of generality $P(y) > 0$, $Q(y) > 0$ for each $y \in (\alpha, \beta)$, $\alpha < \beta$; $\alpha, \beta \in \mathbf{R}^1$. Eventually $P(\alpha)Q(\alpha) = 0$, $P(\beta)Q(\beta) = 0$.

We suppose that:

$$\begin{vmatrix} P(y) = c_1(y - \alpha)^{m_1}(1 + o(1)), m_1 \geq 0, c_1 > 0, y > \alpha, y \to \alpha \\ Q(y) = c_2(y - \alpha)^{m_2}(1 + o(1)), m_2 \geq 0, c_2 > 0, y > \alpha, y \to \alpha \end{vmatrix} \qquad (3.16)$$

$$\left|\begin{array}{l} P(y) = c_3(\beta - y)^{n_1}(1 + o(1)), n_1 \ge 0, c_3 > 0, y < \beta, y \to \beta \\ Q(y) = c_4(\beta - y)^{n_2}(1 + o(1)), n_2 \ge 0, c_4 > 0, y < \beta, y \to \beta \end{array}\right. \quad (3.17)$$

We will concentrate here on the case:

$$\left|\begin{array}{l} y' = \sqrt{\frac{P(y)}{Q(y)}} \\ y(x_0) = y_0 \in (\alpha, \beta), \text{i.e. } y \in (\alpha, \beta) \end{array}\right. \quad (3.18)$$

and consequently, $y(x)$ is strictly monotonically increasing solution of (3.15). Certainly, we will use in many cases even continuation of the solution and when it is possible — its periodic continuation on the real line $\mathbf{R}_x^1$. The continuations satisfy (3.15). They are continuous functions but are non smooth at the peak and cusp points.

Denote now

$$F(y) = \int_{y_0}^{y} \sqrt{\frac{Q(\lambda)}{P(\lambda)}} d\lambda, \alpha < y < \beta.$$

Evidently $F'(y) > 0$, $F(y_0) = 0$, $F(y) < 0$ for $\alpha < y < y_0$ and $F(y) > 0$ for $\beta > y > y_0$. Certainly, $x - x_0 = F(y)$, $y = y(x)$.

There are 4 different cases concerning the behavior of $F(y)$ for $y \to \alpha$ and $y \to \beta$. We give them below.

Thus,

$$(A) \quad 1) \quad \lim_{y \to \alpha} F(y) = \tilde{c} < 0 \iff \frac{m_2 - m_1}{2} + 1 > 0$$

$$(A) \quad 2) \quad \lim_{y \to \alpha} F(y) = -\infty \iff \frac{m_2 - m_1}{2} + 1 \le 0$$

$$(A) \quad 3) \quad \lim_{y \to \beta} F(y) = d > 0 \iff \frac{n_2 - n_1}{2} + 1 > 0$$

$$(A) \quad 4) \quad \lim_{y \to \beta} F(y) = \infty \iff \frac{n_2 - n_1}{2} + 1 \le 0.$$

In a natural way we come to the following subcases of the cases (A) 1)–(A) 4):

$$1a) \quad m_2 - m_1 > 0 \Rightarrow F'(\alpha) = 0$$
$$1b) \quad m_2 - m_1 = 0 \Rightarrow F'(\alpha) > 0$$
$$1c) \quad -2 < m_2 - m_1 < 0 \Rightarrow F'(\alpha) = +\infty$$
$$2) \quad m_2 - m_1 \le -2 \Rightarrow F'(\alpha) = +\infty$$
$$3d) \quad n_2 - n_1 > 0 \Rightarrow F'(\beta) = 0$$
$$3e) \quad n_2 = n_1 \Rightarrow F'(\beta) > 0$$
$$3f) \quad -2 < n_2 - n_1 < 0 \Rightarrow F'(\beta) = +\infty$$
$$4) \quad n_2 - n_1 \le -2 \Rightarrow F'(\beta) = +\infty.$$

This is the first central result of Chapter III.

Theorem 3.1. *(i) Suppose that (A) 1), (A) 3) hold and denote by $\frac{T}{2} = \int_{\alpha}^{\beta} \sqrt{\frac{Q(\lambda)}{P(\lambda)}} d\lambda > 0$. Then (3.15) possesses a strictly monotonically increasing solution for $x \in [0, \frac{T}{2}]$. Moreover, in the case (A) 1) $y \approx \alpha + [(\frac{m_2 - m_1}{2} + 1)\sqrt{\frac{c_1}{c_2}} x]^{\overline{1 + \frac{1}{\frac{m_2 - m_1}{2}}}}$, $x \approx 0$, $x > 0$, while in the case (A) 3) $y \approx \beta - [(\frac{T}{2} - x)\sqrt{\frac{c_3}{c_4}}(\frac{n_2 - n_1}{2} + 1)]^{\overline{1 + \frac{1}{\frac{n_2 - n_1}{2}}}}$, $x \approx \frac{T}{2}$, $x < \frac{T}{2}$.*

(ii) Suppose that (A) 2) and (A) 3) hold.

Then (3.15) possesses a strictly monotonically increasing solution on the half line $(-\infty, 0]$. Moreover, the solution has a horizontal asymptote at $y = \alpha$, while $y \approx \beta - [-(1 + \frac{n_2 - n_1}{2})\sqrt{\frac{c_3}{c_4}} x]^{\overline{1 + \frac{1}{\frac{n_2 - n_1}{2}}}}$, $x \approx 0$, $x < 0$.

(iii) Assume that (A) 2), (A) 4) hold. Then (3.15) has a kink type solution, i.e. strictly monotonically increasing solution with two horizontal asymptotes at $y = \alpha$ and $y = \beta$.

(iv) Assume that (A) 1), (A) 4) hold. Then (3.15) has a monotonically increasing solution in the interval $[0, +\infty)$. Moreover, its behavior for $x \to 0$ is given in (i), while it has a horizontal asymptote at $y = \beta$.

Geometrical interpretation of several cases of Theorem 1 is given on Figs. 3.1, 3.2. We illustrate there peakon y_2 and cuspon y_1 with the same vertex.

As it can be seen easily in all the cases except the case 2), 4) the solution $y(x)$ can be prolonged in an even way, i.e. $y(-x) = y(x)$ and in the cases 1),3) periodically with period T, i.e. $y(x+T) = y(x)$, $\forall x$. The continuation of $y(x)$ satisfies (3.15) but it is non-smooth at the cusp and peak points. Suppose that $\alpha = 0$ and denote by $z(x)$ the restriction of $y(x)$ on the interval $[0, T]$ prolonged at $\mathbf{R}^1 \setminus [0, T]$ as 0. Then we shall say that $z(x)$ is compactly supported (shortly compacton if it it does not lead to confusions) in the cases 1), 3). Assume that the function $y = w(x)$, $a - \varepsilon < x < a + \varepsilon$, $\varepsilon > 0$ has a peak or a cusp point at $b = w(a)$ and a local maximum (minimum) in the same point. We shall say then as in Chapter I that $y = w(x)$ is (locally) a peakon, respectively cuspon at (a, b). Therefore, $z(x)$ is a compacton- cuspon in the cases 1a), 3d); 1b); 3d) and 1c), 3d), while it is a compacton-peakon in the cases 1a), 3e); 1b), 3e) and 1c), 3e). Consider now the case 1), 4) and let $z(x)$ stand for the continuation of $y(x)$ in an even way, i.e. $y(-x) = y(x)$, $\forall x \in \mathbf{R}^1$. The continuation satisfies (3.15). Put $\beta = 0$. Then we shall say that $z(x)$ is a cuspon-soliton in the case 1a), 4),

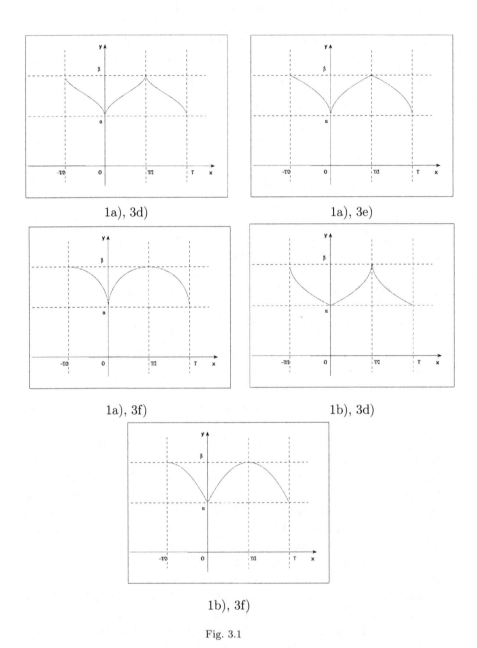

1a), 3d)   1a), 3e)

1a), 3f)   1b), 3d)

1b), 3f)

Fig. 3.1

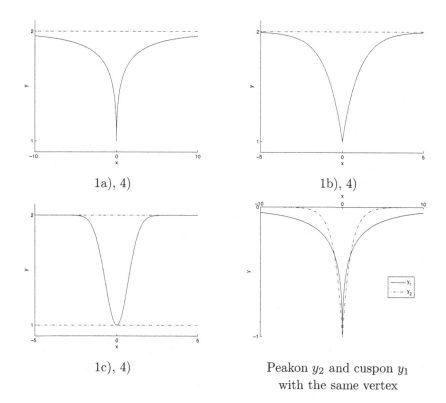

1a), 4)   1b), 4)

1c), 4)   Peakon y_2 and cuspon y_1
                                 with the same vertex

Fig. 3.2

peakon-soliton in the case 1b),4) and soliton in the case 1c), 4). In many papers and books solitons(-peakons,-cuspons) are defined for arbitrary β and not only for $\beta = 0$.

3. Consider now the equation (3.15) supposing that $P(y) > 0$, $Q(y) > 0$ for $y > \alpha$ and (A) 1), (A) 2) hold.

We assume that at infinity

$$P(y) \approx c_5 y^{p_1}, y \to \infty; p_1 \in \mathbf{N} \cup \{0\}, c_5 > 0,$$

$$Q(y) \approx c_6 y^{p_2}, y \to \infty; p_2 \in \mathbf{N} \cup \{0\}, c_6 > 0.$$

Therefore, 2 different cases appear, namely

$$(B) \quad \begin{cases} 5) \ lim_{y \to \infty} F(y) = e > 0 \iff \frac{p_2 - p_1}{2} + 1 < 0 \\ 6) \ lim_{y \to \infty} F(y) = +\infty \iff \frac{p_2 - p_1}{2} + 1 \geq 0. \end{cases}$$

We propose now the second central result of this Chapter.

Theorem 3.2. *(v) Suppose that (A) 1) and (B) 5) are satisfied and denote by $\frac{S}{2} = \int_\alpha^\infty \sqrt{\frac{Q(\lambda)}{P(\lambda)}} d\lambda > 0$, $S < \infty$. Then (3.15) possesses a strictly monotonically increasing solution for $x \in [0, \frac{S}{2})$, having a vertical asymptote at $x = \frac{S}{2}$. Moreover, in the case (A) 1a) it forms a cusp at the point $x = 0$, $y = \alpha$, in the case (A) 1b) it forms a peak at $x = 0$, $y = \alpha$, while in the case (A) 1c) it has a horizontal tangent at $x = 0$, $y = \alpha$.*

(vi) Suppose that (A) 1), (B) 6) hold. Then the solution $y(x)$ exists on the interval $[0, \infty)$ and $\lim_{x \to \infty} y(x) = +\infty$. The behavior of $y(x)$ near the point $x = 0$, $y = \alpha$ for $x > 0$ is the same as in (v).

(vii) Suppose that (A) 2), (B) 5) are valid. Then (3.15) has a strictly monotonically increasing solution y on the interval $(-\infty, 0)$. Moreover, $y(x)$ has a horizontal asymptote at $y = \alpha$ and a vertical asymptote at $x = 0$.

(viii) Assume that (A) 2), (B) 6) are satisfied. Then (3.15) possesses a strictly monotonically increasing solution for each $x \in (-\infty, +\infty)$. Moreover, $y(x)$ has a horizontal asymptote at $y = \alpha$, while $\lim_{x \to \infty} y(x) = +\infty$.

An interesting solution can be constructed in case (viii) as a combination of two solutions: $y(x)$, $x \in \mathbf{R}^1$ and $y(-x)$, $x \in \mathbf{R}^1$ of (3.15). More precisely, we denote by $y_1 = y(x)$ and by $y_2 = y(-x)$.

For the geometrical interpretation of several cases of Theorem 3.2 see Fig. 3.3.

Combining the solutions $y_2 = y(-x)$ and $y_1 = y(x)$, $x \leq 0$ considered in the case 2), 6) and supposing $\alpha = 0$ we obtain a peakon-soliton solution of (3.15), 2), 6).

3.2   Proof of Theorems 3.1 and 3.2

1. Let $y \in [\alpha, \beta]$. Then $F(y) \in [F(\alpha), F(\beta)]$, $F(\alpha) \geq -\infty$, $F(\beta) \leq \infty$ and therefore $x \in [x_0 + F(\alpha), x_0 + F(\beta)]$.

In the case (A) 1) we take $x_0 = -F(\alpha) = -\tilde{c} > 0 \Rightarrow x \in [0, F(\beta) - F(\alpha)] = [0, \int_\alpha^\beta \sqrt{\frac{Q(\lambda)}{P(\lambda)}} d\lambda]$. Put $\int_\alpha^\beta \sqrt{\frac{Q(\lambda)}{P(\lambda)}} = \frac{T}{2}$. Evidently, in the case (A) 1), (A) 3) we have that $0 < T < \infty$, i.e. $x \in [0, \frac{T}{2}]$, while (A) 1), (A) 4) implies that $T = \infty$, i.e. $x \in [0, \infty)$. Obviously, in the case (A) 1)

$$ x - x_0 = F(y) \Rightarrow 0 \leq x = \int_\alpha^y \sqrt{\frac{Q(\lambda)}{P(\lambda)}} d\lambda \leq \frac{T}{2}. $$

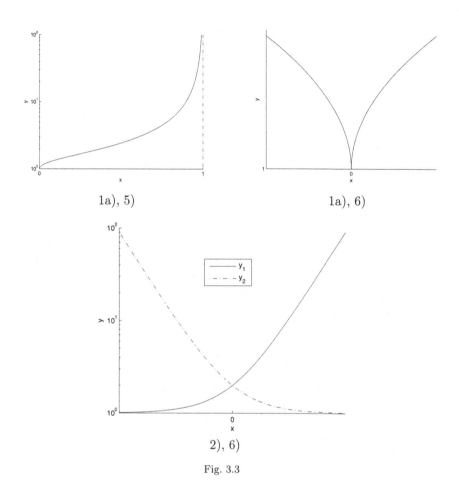

1a), 5)   1a), 6)

2), 6)

Fig. 3.3

Therefore, $x \approx \sqrt{\frac{c_2}{c_1}} \frac{1}{\frac{m_2-m_1}{2}+1}(y-\alpha)^{\frac{m_2-m_1}{2}+1}$, $y \to \alpha$, $y > \alpha$ and conse-

quently, $y(x) \approx \alpha + [(\frac{m_2-m_1}{2}+1)\sqrt{\frac{c_1}{c_2}}x]^{\frac{1}{1+\frac{m_2-m_1}{2}}}$, $x \to 0$, $x > 0$.

In the case (A) 1), (A) 3) we have that $x - \frac{T}{2} = -\int_y^\beta \sqrt{\frac{Q(\lambda)}{P(\lambda)}}d\lambda$, i.e.

$0 \le \frac{T}{2} - x \approx \sqrt{\frac{c_4}{c_3}} \frac{1}{\frac{n_2-n_1}{2}+1}(\beta-y)^{\frac{n_2-n_1}{2}+1}$, $y \to \beta$, $y < \beta$. Thus, $y \approx$

$\beta - [(\frac{T}{2} - x)\sqrt{\frac{c_3}{c_4}}(\frac{n_2-n_1}{2}+1)]^{\frac{1}{1+\frac{n_2-n_1}{2}}}$, $x \to \frac{T}{2}$, $x < \frac{T}{2}$. This way we

complete the proof of Theorem 3.1.

2. Suppose now that $F(\alpha) = -\infty$, $0 < F(\beta) < \infty$ (case 2), 3)). We take $x_0 = -F(\beta) = -d < 0$ and conclude that $x \in (-\infty, 0]$,

$x + F(\beta) = \int_{y_0}^{y} \sqrt{\frac{Q(\lambda)}{P(\lambda)}} d\lambda \Rightarrow x = \int_{\beta}^{y} \sqrt{\frac{Q(\lambda)}{P(\lambda)}} d\lambda \leq 0.$ Consequently,

$x \approx -\sqrt{\frac{c_4}{c_3}} \frac{1}{\frac{n_2-n_1}{2}+1} (\beta - y)^{\frac{n_2-n_1}{2}+1}$, $y \to \beta$, $y > \beta$, i.e. $y \approx \beta - [-(1 +$

$\frac{n_2-n_1}{2}) \sqrt{\frac{c_3}{c_4}} x]^{1+\frac{1}{\frac{n_2-n_1}{2}}}$, $x \to 0$, $x < 0$.

On the other hand, $y(x) \to_{x \to -\infty} \alpha$ as $\int_{\beta}^{y} \sqrt{\frac{Q(\lambda)}{P(\lambda)}} d\lambda \to -\infty$ for $y \to \alpha$. So we complete the proof of Theorem 1, (ii). The proofs of (iii), (iv) are similar to the previous ones and we omit them.

3. Proof of Theorem 3.2. In the case 1), 5) we have that $y \in (\alpha, \infty) \Rightarrow x \in [F(\alpha) + x_0, x_0 + e)$. Taking $x_0 = -F(\alpha) = -\tilde{c} > 0$ and having in mind that $F(\infty) = e > 0$ we conclude that $y \in (\alpha, \infty) \Rightarrow x \in [0, \frac{S}{2})$, where $0 < \frac{S}{2} = \int_{\alpha}^{\infty} \sqrt{\frac{Q(\lambda)}{P(\lambda)}} d\lambda < \infty$ and more precisely, $x = x_0 + F(y) = \int_{\alpha}^{y} \sqrt{\frac{Q(\lambda)}{P(\lambda)}} d\lambda$, $y \in [\alpha, \infty)$; $y \to +\infty \Rightarrow x \to \frac{S}{2}$. We obtain that the inverse function $y = y(x)$ is defined on the interval $[0, \frac{S}{2})$ and has a vertical asymptote at the point $x = \frac{S}{2}$. Combining this observation and Theorem 3.1 (i) we prove Theorem 3.2 (v). In the case 1), 6) we take again $x_0 = -F(\alpha)$ but then $F(\infty) = \infty$ and therefore $y \in [\alpha, \infty) \Rightarrow x \in [0, \infty)$. Moreover, $x = \int_{\alpha}^{y} \sqrt{\frac{Q(\lambda)}{P(\lambda)}} d\lambda$. Thus, the inverse function $y = y(x) \to \infty$ for $x \to \infty$, while $y(0) = \alpha$. Theorem 3.2 (vi) is shown.

In the case 2), 5) we have that $y \in (\alpha, \infty) \Rightarrow F(y) \in (-\infty, e)$, $e = F(\infty)$. Because of this we take $x_0 = -e = -\int_{y_0}^{\infty} \sqrt{\frac{Q(\lambda)}{P(\lambda)}} d\lambda < 0$ and we conclude that $y \in (\alpha, \infty) \Rightarrow x = x_0 + F(y) \in (-\infty, 0)$, $x = \int_{+\infty}^{y} \sqrt{\frac{Q(\lambda)}{P(\lambda)}} d\lambda$. The inverse function $y = y(x)$ is defined for $x \in (-\infty, 0)$ and possesses a horizontal asymptote $y = \alpha$ and a vertical asymptote at $x = 0$ because $y(x) \to \alpha$ for $x \to -\infty$, $y(x) \to \infty$ for $x \to 0$, $x < 0$.

Case 2), 6) (i.e. (viii)) is simpler to deal with. To see this we observe that $y \in (\alpha, \infty) \Rightarrow F(y) \in (-\infty, \infty)$, i.e. $x = x_0 + F(y) \in (-\infty, \infty)$ defines on $\mathbf{R}^1$ a monotonically increasing function $y(x)$ with horizontal asymptote at $y = \alpha(y \to \alpha \Rightarrow x \to -\infty)$, $\lim_{x \to \infty} y(x) = \infty$. The initial data x_0, $y_0 > \alpha$ are arbitrary. One can easily guess that the function $z(x) = y(-x)$, $x \in \mathbf{R}^1$ satisfies (3.15). Define now the even solution $w(x)$ of (3.15) by the formula: $w(x) = \begin{cases} y(x), & x \leq 0 \\ y(-x), & x \geq 0 \end{cases}$. Evidently, $w(x)$ is a peakon with a peak at $x = 0$, $y(0)$ and $\lim_{x \to \pm\infty} w(x) = \alpha$. This way everything is proved (see Fig. 3.3; 2), 6) for the geometrical visualisation).

We shall illustrate Theorem 3.1 by the following example of CH equation considered in [89].

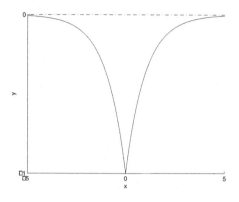

Fig. 3.4

Example 1. Let

$$u_t + 2k^2 u_x + 3uu_x - u_{xxt} = \left[\frac{u_x^2}{2} + uu_{xx}\right]_x. \qquad (3.19)$$

Putting $u = \varphi(x + ct)$, $c = const \neq 0$, $k = const$, $\xi = x + ct$ we obtain

$$C_1\varphi + (c + 2k^2)\varphi^2 + \varphi^3 - C_2 = (\varphi')^2(c + \varphi) \qquad (3.20)$$

and C_1, C_2 are arbitrary constants.

If we are looking for a soliton solutions of (3.19) we must take $\varphi(\infty) = \varphi'(\infty) \Rightarrow C_2 = 0$.

Thus, under the additional assumption $C_1 = 0$ (3.20) takes the form:

$$\varphi^2[(c + 2k^2) + \varphi] = (\varphi')^2(c + \varphi). \qquad (3.21)$$

Denote by $A = c + 2k^2$ and suppose that $A > c > 0$. Obviously, we will investigate (3.21) for $0 > \varphi > -c > -A$, i.e. $\xi = F(\varphi) = \int_{-c}^{\varphi} \frac{\sqrt{c + \lambda} d\lambda}{|\lambda|\sqrt{A + \lambda}} = \int_{-c}^{\varphi} \frac{c + \lambda}{|\lambda| X^{1/2}} d\lambda$ is the strictly monotonically increasing solution of (3.21) passing through the point $\xi = 0$, $\varphi(0) = -c$, where $X(\lambda) = \lambda^2 + (c + A)\lambda + Ac$ and $X > 0$ for $\lambda \in (-c, 0)$. Evidently, $\varphi \in [-c, 0) \Rightarrow \xi = F(\varphi) \in [0, \infty)$. Geometrically we have (see Fig. 3.4):

Consequently, (3.21) possesses a solution of the type soliton-cuspon. It is rather simple to compute the definite integral $F(\varphi)$. In fact $F(\varphi) = -cI - II$, $\varphi \in [-c, 0)$ and $I = \int_{-c}^{\varphi} \frac{d\lambda}{\lambda X^{1/2}}$, $II = \int_{-c}^{\varphi} \frac{d\lambda}{\sqrt{X(\lambda)}}$. Therefore (see [45]),

$$I = -\frac{1}{\sqrt{Ac}} \ln\left|\frac{2\sqrt{Ac}X^{1/2} + 2Ac + \lambda(A + c)}{\lambda}\right|\Big|_{\lambda = -c}^{\lambda = \varphi},$$

$$II = \ln|2X^{1/2} + 2\lambda + A + c|\Big|_{\lambda = -c}^{\lambda = \varphi}.$$

The observation that $-c < \lambda < 0 \Rightarrow 2\lambda + A + c > 0$, $(A + c)\lambda + 2Ac > c(A - c) > 0$ gives us that for $\xi \geq 0$, $-c \leq \varphi < 0$

$$\frac{[2\sqrt{Ac}\sqrt{X(\varphi)} + (A+c)\varphi + 2Ac]\sqrt{\frac{c}{A}}}{|\varphi|^{\sqrt{\frac{c}{A}}}(2\sqrt{X(\varphi)} + 2\varphi + A + c)} = e^\xi \frac{[c(A-c)]\sqrt{\frac{c}{A}}}{c^{\sqrt{\frac{c}{A}}}(A-c)} = e^\xi (A-c)^{\sqrt{\frac{c}{A}}-1}.$$
$$(3.22)$$

Proposition 3.1. *For each real $k \neq 0$ in (3.19) and $\frac{p}{q} \in Q \cap (0,1)$ there exists a velocity $c > 0$ and such that $1 > \sqrt{\frac{c}{A}} = \frac{p}{q} \in Q$, i.e. $c = \frac{r^2 k^2}{1-r^2}$, $r = \frac{p}{q}$, while the function $\varphi(\xi)$ given by (3.22), $\xi \geq 0$, $0 > \varphi > -c$, satisfies an algebraic equation of order $2(p+q)$ namely, $P_{2(p+q)}(\varphi) = 0$, whose coefficients depend linearly on $e^{q\xi}$, $e^{2q\xi}$.*

We omit the elementary algebraic proof of this assertion.

Remark 3.1. (see [89]) Put $C_1 = 0$ and $k = 0$ in (3.20). Then the monotonically increasing solution of (3.21) with $\varphi_0(0) = -c$, $-c \leq \varphi_0 < 0$ is given by $\varphi' = |\varphi|$, $\varphi(0) = -c$, i.e. $\varphi_0(\xi) = -ce^{-\xi}$, $\xi \geq 0$ and we have a soliton-peakon solution with peak at $\xi = 0$, $\varphi(0) = -c$, $\varphi_0 = -ce^{-|\xi|}$, $\xi \in \mathbf{R}^1$.

Denote by $\varphi_k(\xi)$ the soliton-cuspon solution of the CH equation (3.21) in the special case $C_1 = 0$. We know that $\xi \in [0,\infty)$, $\varphi_k(0) = 1$, $0 > \varphi_k(\xi) > -c$ for $\xi \neq 0$ and $\lim_{\xi\to\infty}\varphi_k(\xi) = 0$. Fix now the point $\xi > 0$ and let $k > l > 0$. Then there are no difficulties to verify that $0 > \varphi_k(\xi) > \varphi_l(\xi) > -c$. In fact,

$$\int_{-c}^{\varphi_l(\xi)} \frac{(c+\lambda)d\lambda}{|\lambda|\sqrt{(c+2l^2+\lambda)(c+\lambda)}} = \xi = \int_{-c}^{\varphi_k(\xi)} \frac{(c+\lambda)d\lambda}{|\lambda|\sqrt{(c+2k^2+\lambda)(c+\lambda)}}$$
$$< \int_{-c}^{\varphi_k(\xi)} \frac{(c+\lambda)d\lambda}{|\lambda|\sqrt{(c+2l^2+\lambda)(c+\lambda)}}. \qquad (3.23)$$

The strict monotonicity of the defined above function $F(\varphi)$ for $-c < \varphi < 0$ implies that $\varphi_k(\xi) > \varphi_l(\xi)$. Consequently, there exists $\lim_{k\to 0}\varphi_k(\xi) = l_0(\xi) \geq -c$. Letting $k \to 0$ in (3.23) we obtain:

$$\int_{-c}^{l_0(\xi)} \frac{d\lambda}{|\lambda|} = \xi \Rightarrow l_0(\xi) = -ce^{-\xi} \Rightarrow l_0 \equiv \varphi_0.$$

Conclusion. Consider the equation (3.21) with $k > 0$ small parameter. The corresponding sequence φ_k of soliton-cuspon solutions of (3.21) tends pointwisely to the soliton-peakon solution φ_0 of (21), where $k = 0$ (see Fig. 3.5).

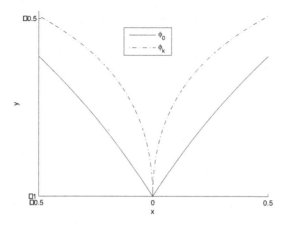

Fig. 3.5

3.3 Elliptic functions and applications to several equations of mathematical physics

1. We shall begin with a small excursion in the classical theory of elliptic functions and elliptic integrals of first, second and third kind in $\mathbf{R}^1$ and in the form of Legendre ([48], [2], [100], [21]).

Thus, consider the following autonomous system of ODE:

$$f' = gh, f = f(x), g = g(x), h = h(x), x \in \mathbf{R}^1$$
$$g' = -hf, \tag{3.24}$$
$$h' = -k^2 fg, f(0) = 0, g(0) = 1, h(0) = 1,$$

where k is a parameter. We shall take $|k| \leq 1$, $k \in \mathbf{R}^1$.

It is evident that $ff' + gg' = 0 \iff f^2(x) + g^2(x) = 1$, i.e. $|f| \leq 1$, $|g| \leq 1$. Moreover, $k^2 ff' + hh' = 0 \iff k^2 f^2(x) + h^2(x) \equiv 1$, i.e. $h^2 = 1 - k^2 f(x) \Rightarrow |h| \leq 1$. Certainly, $k^2 g^2(x) - h^2(x) \equiv k^2 - 1$. It is well known from the theory of ODE that $|f| \leq 1$, $|g| \leq 1$, $|h| \leq 1$ imply that the Cauchy problem (3.24) possesses a unique solution for each $x \in \mathbf{R}^1$ and this solution is an analytic (real) function of x. Following Jacobi, we denote: $f(x) = sn(x, k)$ and call it elliptic sine, $g(x) = cn(x, k)$ and call it elliptic cosine and $h(x) = dn(x, k)$ and call it delta elliptic function. Having in mind the three first integrals of the system (3.24) obtained above, we get:

$$(f')^2 = g^2 h^2 = (1 - f^2)(1 - k^2 f^2)$$
$$(g')^2 = (1 - g^2)(k^2 g^2 + 1 - k^2) \qquad (3.25)$$
$$(h')^2 = -(1 - h^2)(-h^2 + 1 - k^2), f(0) = 0, g(0) = 1, h(0) = 1.$$

Therefore,

$$F(f) = \int_0^f \frac{d\lambda}{\sqrt{(1 - \lambda^2)(1 - k^2\lambda^2)}} = x, 0 \leq k^2 \leq 1, 0 \leq f \leq 1,$$

$$G(\varphi) = \int_1^g \frac{d\lambda}{\sqrt{(1 - \lambda^2)(k^2\lambda^2 + 1 - k^2)}}, 0 \leq g \leq 1$$

and similar expression can be found for h.

The function $F(f)$ is defined and strictly monotonically increasing on the interval $f \in [0, 1]$, $0 \leq k^2 < 1$, $F'(0) = 1$, $F'(1) = +\infty$, i.e. the inverse function $f = F^{-1}(x)$ is defined and monotonically increasing on the interval $x \in [0, K(k)]$, where $K(k) = \int_0^1 \frac{d\lambda}{\sqrt{(1-\lambda^2)(1-k^2\lambda^2)}} < \infty$, $f(0) = 0$, $f(K(k)) = 1$, $f'(0) = 1$, $f'(1) = 0$, $0 \leq k^2 < 1$, while $K(1) = +\infty$. We shall continue $f(x)$ on the interval $[-K(k), 0]$ in an odd way, i.e. $f(-x) = -f(x)$, $x \in [0, K(k)]$ and $f(x + K(k)) = f(K(k) - x)$ for $x \in [0, K(k)]$; $f(-x) = -f(x)$, $\forall x \in [-2K(k), 0]$. Then we continue $f(x)$ in a periodic way with period $4K(k)$ along $\mathbf{R}^1$ (see Fig. 3.6). Due to the unicity of the solution of (3.24) and the smoothness of $f(x)$ we obtain $sn(x, k) = f(x)$, $\forall x \in \mathbf{R}^1$. Shortly we shall write $sn\, x = f(x)$.

The integral $K(k)$ is called complete elliptic integral of the first kind. Evidently, $sn(x, k) = 0 \iff x = 2jK(k)$, $j \in \mathbf{Z}$; $sn(x, 0) = sin\, x$, $cn(x, 0) = cos\, x$. In a similar way we can prove that the function $cn(x, k)$ is analytic and periodic with period $4K(k)$, $cn(x, k) = 0 \iff x = (2j + 1)K(k)$, $j \in \mathbf{Z}$. The function $dn(x, k)$ is periodic with period $2K(k)$ as $h^2(x) = k^2 g^2(x) - k^2 + 1 \geq 1 - k^2$, $1 \geq h(x) \geq \sqrt{1 - k^2}$. In other words, for each $k \in [0, 1)$ the periodic function $h(x)$ oscillates between $0 < \sqrt{1 - k^2}$ and 1. It is easy to see that

$$K(k) = \int_0^1 \frac{d\lambda}{\sqrt{(1 - \lambda^2)(1 - k^2\lambda^2)}} = \int_0^{\frac{\pi}{2}} \frac{d\alpha}{\sqrt{1 - k^2 sin^2\alpha}}, |k| < 1, \quad (3.26)$$

respectively, $F(f) = x$ takes the form:

$$x = \int_0^\varphi \frac{d\alpha}{\sqrt{1 - k^2 sin^2\alpha}}, 0 \leq \varphi \leq \frac{\pi}{2}, \qquad (3.27)$$

where $sin\, \varphi = f$.

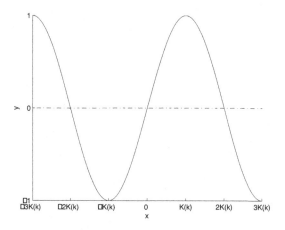

Fig. 3.6

Therefore,

$$sn(x, k) = f(x) = sin\,\varphi, \qquad (3.28)$$

$\varphi = am\ x$ being the inverse function of the strictly monotonically increasing function (3.27).

Consider now the case $|k| = 1$. Then

$$x = \int_0^f \frac{d\lambda}{1 - \lambda^2} = \frac{1}{2}ln\frac{1 + f}{1 - f} \Rightarrow f(x) = th\ x$$

(hyperbolic tangent of x), i.e. $sn(x, 1) = th\ x$ and evidently, $cn(x, 1) = dn(x, 1) = \frac{1}{ch\ x}$ as

$$\int_1^g \frac{d\lambda}{\lambda\sqrt{1 - \lambda^2}} = x \Rightarrow g(x) = \frac{1}{ch\ x}, g(0) = 1.$$

From the system (3.24) we have that $[sn(x, k)]' = cn(x, k)dn(x, k)$, $[cn(x, k)]' = -sn(x, k)dn(x, k)$ and $[dn(x, k)]' = -k^2sn(x, k)cn(x, k)$. Having in mind the definition of $f(x) = sn(x, k)$, $x \in (-K(k), K(k))$, we have that it is invertible and $x = F(f) \Rightarrow \frac{dx}{df} = \frac{1}{\sqrt{(1 - f^2)(1 - k^2 f^2)}}$, $|f| < 1$, where $x = sn^{-1}(f, k)$. Evidently, $x = \int_0^{sn(x,k)} \frac{d\lambda}{\sqrt{(1 - \lambda^2)(1 - k^2 \lambda^2)}}$ and the inverse function $x = sn^{-1}(f, k)$, $|f| < 1$, of $f = sn(x, k)$ is given by $x = sn^{-1}(f, k) = \int_0^f \frac{d\lambda}{\sqrt{(1 - \lambda^2)(1 - k^2 \lambda^2)}}$.

To complete the things we shall define now the elliptic integrals of first, second and third kind, respectively:

$$F(\varphi, k) = \int_0^\varphi \frac{d\alpha}{\sqrt{1 - k^2 sin^2\alpha}} = \int_0^{sin\,\varphi} \frac{d\lambda}{\sqrt{(1 - \lambda^2)(1 - k^2\lambda^2)}}, |k| < 1,$$

$$(3.29)$$

(see (3.27) implying that if $sin\,\varphi = sn(x, k)$ then $F(\varphi, k) = x$),

$$E(\varphi, k) = \int_0^\varphi \sqrt{1 - k^2 sin^2\alpha}\,d\alpha = \int_0^{sin\,\varphi} \frac{\sqrt{1 - k^2\lambda^2}}{\sqrt{1 - \lambda^2}}\,d\lambda, \qquad (3.30)$$

$$\Pi(\varphi, n, k) = \int_0^\varphi \frac{d\alpha}{(1 + n\,sin^2\alpha)\sqrt{1 - k^2 sin^2\alpha}}$$

$$= \int_0^{sin\,\varphi} \frac{d\lambda}{(1 + n\lambda^2)\sqrt{(1 - \lambda^2)(1 - k^2\lambda^2)}}. \qquad (3.31)$$

We shall write as above the complete integrals of first and second kind: $K(k) = F(\frac{\pi}{2}, k)$, $E(K) = E(\frac{\pi}{2}, k)$. Certainly, $F(-\varphi, k) = -F(\varphi, k)$, $E(-\varphi, k) = -E(\varphi, k)$,

$$F(n\pi \pm \varphi, k) = 2\pi K(k) \pm F(\varphi, k), n \in \mathbf{Z},$$

$$E(n\pi \pm \varphi, k) = 2\pi E(k) \pm E(\varphi, k), n \in \mathbf{Z}.$$

Put $k'(k) = \sqrt{1 - k^2}$ and $K'(k) = F(\frac{\pi}{2}, \sqrt{1 - k^2})$, $|k| < 1$. Then (see [2]):

$$sn(x, k) = \frac{2\pi}{kK(k)} \sum_{n=1}^\infty \frac{q^{n-1/2}}{1 - q^{2n-1}} sin\left[(2n - 1)\frac{\pi}{2K(k)}\right],$$

where $q = e^{-\pi\frac{K'(k)}{K(k)}}$ ($q < 1$). This is the Fourier expansion of the periodic function $sn(x, k)$, $|k| < 1$, of course. A similar formula is valid for $cn(x, k)$. This way we have explicit formulas for the analytic functions $sn(x, k)$, $cn(x, k)$. To complete these short notes we point out that $lim_{k\to 0}K(k) = \frac{\pi}{2}$, $K'(k) \approx ln\frac{4}{k}$ for $k \to 0$, $k > 0$ [2].

1. One can easily check that

$$\int cn(x, k)dx = \frac{1}{k}arccos(dn(x, k)), |k| < 1, k \neq 0$$

$$\int dn(x, k)dx = arcsin(sn(x, k)), |k| < 1,$$

$$\int sn(x, k)dx = -\frac{1}{k}arcch\left(\frac{dn(x, k)}{k'}\right), k' = \sqrt{1 - k^2},$$

$|k'| < 1$. We remind of the reader that $2ch\ x = e^x + e^{-x} \geq 2$, the inverse function $Arcch\ x$ of $ch\ x$, is double valued, $Arcch\ x = \pm ln(x + \sqrt{x^2 - 1})$, $x > 1$ and $(Arcch\ x)' = \frac{\pm 1}{\sqrt{x^2 - 1}}$, $x > 1$. By $arcch\ x$ we have denoted an appropriate branch of $Arcch\ x$. The above written formulas are useful in finding the explicit form of the solutions of some equations of mathematical physics as Benney-Luke equation (see [73]).

2. We proposed above an exuberance of formulas for the elliptic functions. Instead of comments on them we prefer to illustrate them by several examples from classical mechanics provided with the corresponding calculations. We shall not enter into details but our aim is to stress the most important technicalities.

Example 2. Consider the ODE $(u')^2 = \Phi(u)$ with separate variables written into integral form: $t - t_0 = \int_{u_0}^{u} \frac{du}{\Phi(u)}$, where $\Phi(u) = \alpha(e_1 - u)(u - e_2)(u - e_3)$, $\alpha = const \neq 0$, $e_3 < e_2 < e_1$ are the real roots of the cubic polynomial $\Phi(u) = 0$; $e_2 < u < e_1 \Rightarrow \Phi(u) > 0$. We take $u_0 = e_1$. After the change $e_1 - u = (e_1 - e_2)w^2$, $w \geq 0$, i.e. $u = e_1 \Rightarrow w = 0$, $e_2 = u \Rightarrow w = 1$, $du = -2(e_1 - e_2)wdw$ we get $(u = e_1 - (e_1 - e_2)w^2)$

$$t - t_0 = \int_{u_0}^{u} \frac{du}{\Phi(u)} = \frac{-2}{\sqrt{e_1 - e_3}} \int_0^w \frac{wdw}{\sqrt{-\alpha}w\sqrt{(1 - w^2)(1 - k^2w^2)}}$$

$$= \frac{-2v}{\sqrt{\alpha(e_3 - e_1)}}, v \in \mathbf{R}^1,$$

where $k^2 = \frac{e_1 - e_2}{e_1 - e_3}$, $0 < k^2 < 1$, $\alpha < 0$. Consequently, $v = -\frac{\sqrt{\alpha(e_3 - e_1)}}{2}(t - t_0)$, $w = sn\ v \Rightarrow$

$$u = e_1 - (e_1 - e_2)sn^2 v = e_1 - (e_1 - e_2)sn^2\left(\frac{\sqrt{\alpha(e_3 - e_1)}}{2}(t - t_0)\right)$$

$$= e_2 + (e_1 - e_2)cn^2\left(\frac{\sqrt{\alpha(e_3 - e_1)}}{2}(t - t_0)\right). \tag{3.32}$$

The formula (3.32) plays an important role in studying the Euler-Poisson equation that describes the movement of a solid body (this is the so called Lagrange case — see [9]). Another application is to the KdV equation

$$u_t + uu_x + Ku_{xxx} = 0, K = const > 0. \tag{3.33}$$

If we are looking for travelling wave solutions $u = u(t, x) = u(\xi)$, $\xi = x - ct$, $c = const$, then the solvability of (3.33) is reduced to the solvability of the ODE with coefficients depending on c : $\frac{du}{d\xi} = \pm(\frac{1}{3K}f(u))^{1/2}$, $f(u) = (u - \gamma)(u - \beta)(\alpha - u)$.

In the special case when $\gamma < \beta < \alpha$

$$u(\xi) = \beta + (\alpha - \beta)cn^2\left(\xi\sqrt{\frac{\alpha - \gamma}{2K}}, k\right), k^2 = \frac{\alpha - \beta}{\alpha - \gamma}.$$

The case $\gamma = \beta < \alpha$ is simpler to deal with. In fact, then $d\xi = \sqrt{3K}\frac{du}{(u-\gamma)\sqrt{\alpha-u}}$, $sech\ x = \frac{1}{ch\ x}$, $ch\ x = \frac{1}{2}(e^x + e^{-x})$, $(Arcsech)' = \frac{-1}{x\sqrt{1-x^2}}$, $0 < x < 1$, $(Arcsech\sqrt{x})' = -\frac{1}{2x\sqrt{1-x}}$, $0 < x < 1$ (see [45]). Thus,

$$u(\xi) = \gamma + (\alpha - \gamma)sech^2\left(\sqrt{\frac{\alpha - \gamma}{12K}}\xi\right).$$

This is a solution called soliton of course [16], [119].

Example 3. Consider the ODE

$$t - t_0 = \int_{u_0}^{u} \frac{du}{\Phi(u)}, \tag{3.34}$$

where $\Phi(u) = (e_1 - u)(e_2 - u)(e_3 - u)(u - e_4)$, $e_1 > e_2 > e_3 > e_4$; $u_0 = e_4$, $e_3 > u > e_4 \Rightarrow \Phi(u) > 0$.

This is the appropriate change of the variable u in expressing the elliptic integral from (3.34) by the Jacobi functions:

$$(e_1 - u) = \left[\frac{1}{e_1 - e_4} + \left(\frac{1}{e_1 - e_3} - \frac{1}{e_1 - e_4}\right)z^2\right]^{-1}, z \geq 0;$$

$u_0 = e_4 \Rightarrow w = 0$.

Put $A = \frac{1}{e_1-e_4} + (\frac{1}{e_1-e_3} - \frac{1}{e_1-e_4})z^2$, $\frac{1}{e_1-e_4} > 0$, $\frac{1}{e_1-e_3} - \frac{1}{e_1-e_4} > 0$, $e_1 - u = \frac{1}{A}$. Evidently, $du = \frac{1}{A^2}(\frac{1}{e_1-e_3} - \frac{1}{e_1-e_4})2zdz = \frac{e_3-e_4}{(e_1-e_3)(e_1-e_4)}\frac{2z}{A^2}dz$,

$e_2 - u = \frac{1}{A} - (e_1 - e_2)$, $e_3 - u = \frac{1}{A} - (e_1 - e_3)$, $u - e_4 = (e_1 - e_4) - \frac{1}{A}$.

Thus,

$$\sqrt{\Phi(u)} = \frac{1}{A^2}\sqrt{(1 - A(e_1 - e_2))(1 - A(e_1 - e_3))(A(e_1 - e_4) - 1)}$$

$$= \frac{1}{A^2}\sqrt{\left(\frac{e_2 - e_4}{e_1 - e_4} - \frac{(e_1 - e_2)(e_3 - e_4)}{(e_1 - e_3)(e_1 - e_4)}z^2\right)\left(\frac{e_3 - e_4}{e_1 - e_4} - \frac{e_3 - e_4}{e_1 - e_4}z^2\right)}$$

$$\times \sqrt{\frac{e_3 - e_4}{e_1 - e_4}}z = \frac{z}{A^2}\frac{(e_3 - e_4)\sqrt{1 - z^2}}{\sqrt{e_1 - e_3}(e_1 - e_4)}\sqrt{e_2 - e_4}\sqrt{1 - k^2z^2},$$

$0 < k^2 = \frac{(e_1-e_2)(e_3-e_4)}{(e_1-e_3)(e_2-e_4)} < 1$. So $\sqrt{(e_1 - e_3)(e_2 - e_4)}(t - t_0) = 2\int_0^z \frac{dz}{\sqrt{(1-z^2)(1-k^2z^2)}}$. Put $\tau = \frac{\sqrt{(e_1-e_3)(e_2-e_4)}}{2}(t - t_0)$.

Then $z = sn\,\tau \Rightarrow u = e_1 - \dfrac{1}{[\frac{1}{e_1-e_4}+\frac{e_3-e_4}{(e_1-e_4)(e_1-e_3)}\,sn^2\tau]}$.

Integral (3.34) appears in studying the movement of a solid body in the Newtonian field of force (see [9]).

3. In the investigation of the so called fluxons see Chapters 8, 9 from [76] the famous sin-Gordon equation ([119], [105], [92]) and its modifications appear, namely

$$\frac{\partial^2 \varphi}{\partial x^2} - \frac{\partial^2 \varphi}{\partial t^2} = sin\,\varphi, \tag{3.35}$$

respectively

$$\frac{\partial^2 \varphi}{\partial x^2} - \frac{\partial^2 \varphi}{\partial t^2} - \Gamma(1 + \varepsilon cos\,\varphi)|\frac{\partial \varphi}{\partial t}|\frac{\partial \varphi}{\partial t} = sin\,\varphi - \gamma, \tag{3.36}$$

γ, $\Gamma = const > 0$, $0 < \varepsilon \ll 1$. We shall look for travelling wave solutions of the constant speed $c > (<)1$. This way from (3.35) we get

$$\frac{d^2\varphi}{d\xi^2} = \frac{sin\,\varphi}{1 - c^2}. \tag{3.37}$$

R. Parmentier pointed out in [76] that this equation has two different from physical point of view solutions — plasma waves corresponding to the swing (oscillation) of the pendulum and fluxon waves corresponding to the rotation of the pendulum. In the first case $c^2 > 1$, (3.37) implies $\frac{1}{2}(\frac{d\varphi}{d\xi})^2 = \frac{1}{c^2-1}cos\,\varphi - \frac{a}{c^2-1}$, $a > 0$, $a = const < 1$. Thus,

$$d\xi = \frac{d\varphi}{\sqrt{\frac{2}{c^2-1}(cos\,\varphi - a)}} \Rightarrow \xi = \sqrt{\frac{c^2-1}{2}}\int_0^\varphi \frac{d\psi}{\sqrt{cos\,\psi - a}}.$$

Put $a = cos\,\alpha$, $0 < \alpha < \frac{\pi}{2} \Rightarrow \xi = \frac{1}{2}\sqrt{c^2-1}\int_0^\varphi \frac{d\psi}{\sqrt{sin^2\frac{\alpha}{2}-sin^2\frac{\psi}{2}}}$ and denote $0 < k^2 = sin^2\frac{\alpha}{2} < 1$, $sin\frac{\psi}{2} = ysin\frac{\alpha}{2} = ky$; $\psi = 0 \Rightarrow y = 0$, $\psi = \varphi \Rightarrow u = \frac{sin\frac{\varphi}{2}}{sin\frac{\alpha}{2}}$.

After this change of the variable in the above integral we obtain:

$$\xi = \sqrt{c^2-1}\int_0^u \frac{dy}{\sqrt{(1-y^2)(1-k^2y^2)}} \Rightarrow u = sn\left(\frac{\xi}{\sqrt{c^2-1}},k\right),$$

$k = sin\frac{\alpha}{2} \Rightarrow sin\frac{\varphi}{2} = sin\frac{\alpha}{2}sn(\frac{\xi}{\sqrt{c^2-1}},k)$, i.e. $\varphi = 2\,arcsin(ksn(\frac{\xi}{\sqrt{c^2-1}},k))$.

As it concerns the fluxon waves ($c^2 < 1$) is a similar way we obtain

$$\varphi = arcsin\left(\pm cn\left(\frac{-\xi}{k\sqrt{1-c^2}},k\right)\right), 0 < k < 1.$$

Solutions of the same type can be found for (3.36) and with $\varepsilon = 0$. Then the travelling wave solution is given by:

$$\frac{d^2\varphi}{d\xi^2} - \Gamma\frac{c^2}{1-c^2}\left(\frac{d\varphi}{d\xi}\right)^2 = \frac{\sin\varphi - \gamma}{1-c^2}. \tag{3.38}$$

The standard change $\frac{d\varphi}{d\xi} = p(\varphi)$ in (3.38) leads to:

$$\frac{1}{2}\frac{d}{d\varphi}(p^2) - \Gamma\frac{c^2}{1-c^2}p^2 = \frac{\sin\varphi - \gamma}{1-c^2},$$

i.e. we have obtained a linear first order ODE with respect to p^2 and with independent variable φ. Then we must solve the ODE with separate variables $(\frac{d\varphi}{d\xi})^2 = F(\varphi)$ etc.

Answer: $\varphi = arcsin\,\gamma_0 + 2\,arcsin(cn(\frac{1}{k}(\frac{\gamma_0}{2\Gamma c^2})^{\frac{1}{2}}(\xi - \xi_0), k))$, where $k = (\frac{2\gamma_0}{\gamma+\gamma_0})^{1/2}$, $\gamma_0 = \frac{2\Gamma c^2}{((1-c^2)^2+4\Gamma^2 c^2)^{1/2}}$.

This is a travelling wave solution of (3.35) in a very special case:

$$\frac{1}{2}(c^2 - 1)\left(\frac{d\varphi}{d\xi}\right)^2 = \cos\varphi - 1,$$

i.e. for $c^2 < 1$ we have

$$\frac{1}{2}(c^2 - 1)\left(\frac{d\varphi}{d\xi}\right)^2 = -2\sin^2\frac{\varphi}{2} \Rightarrow \frac{d\varphi}{d\xi} = \pm\frac{2}{\sqrt{1-c^2}}\sin\frac{\varphi}{2}.$$

Having in mind that $\int \frac{d\lambda}{\sin\lambda} = ln|tg\frac{\lambda}{2}| + c$, $c = const$ we get that $tg\frac{\varphi}{4} = \pm e^{\pm\frac{\xi-\xi_0}{\sqrt{1-c^2}}}$.

The solutions

$$\varphi_\pm = \pm 4\,arctg\,e^{\pm\frac{\xi-\xi_0}{\sqrt{1-c^2}}} \tag{3.39}$$

are describing geometrically loops ξ varying from $\xi = -\infty$ to $\xi = +\infty$. If we take in (3.39) in both cases the sign " $+$ " we obtain a positive loop, i.e. $\xi \in (-\infty, \infty) \Rightarrow \varphi \in (0, 2\pi)$, while if we take "$-$" in both cases in (3.39) we have again positive loop but $\xi \in (-\infty, +\infty) \Rightarrow \varphi \in (-2\pi, 0)$. The opposite signs in (3.39) give negative loops. We shall use a physical terminology: positive loops are called fluxons and negative loops are called antifluxons.

We are interesting now in finding solutions of (3.35) written in the form (Perring-Skyrme [91]), Lamb [67])

$$\psi = f(x)g(t) = tg\frac{\varphi}{4}, \tag{3.40}$$

i.e. $\varphi = 4\,arctg(f(x)g(t))$.

Evidently, $\varphi_{tt} = \frac{4fg''(1+f^2g^2)-8f^3g(g')^2}{1+\psi^2}$, $\varphi_{xx} = \frac{4g[f''(1+\psi^2)-2fg^2(f')^2]}{1+\psi^2}$.

The trigonometric formulas $sin\,4\gamma = 4cos\,\gamma(sin\,\gamma - 2\,sin^3\gamma)$, $sin\,\alpha = \frac{tg\,\alpha}{\pm\sqrt{1+tg^2\alpha}}$, $cos\,\alpha = \frac{1}{\pm\sqrt{1+tg^2\alpha}}$ imply that after the change (3.40) the equation (3.35) can be rewritten as:

$$(1 + f^2g^2)(fg'' - gf'') + 2fg(g^2(f')^2 - f^2(g')^2) + fg(1 - f^2g^2) = 0. \quad (3.41)$$

Having in mind that $f \equiv f(x)$, $g \equiv g(t)$ we shall look for f and g as solutions of the following ODE with separate variables:

$$(f')^2(x) = P_1(f(x)), P_1 = a_0 f^4 + a_1 f^3 + a_2 f^2 + a_3 f + a_4, \quad (3.42)$$
$$(g')^2(t) = P_2(g(t)), P_2 = b_0 g^4 + b_1 g^3 + b_2 g^2 + b_3 g + b_4, \quad (3.43)$$

a_i, b_j being real constants.

Certainly, (3.42), (3.43) are satisfied by appropriate elliptic functions and the coefficients of the polynomials P_1, P_2 are unknown. Evidently, $f''(x) = \frac{1}{2}\frac{dP_1}{df}$, $g''(t) = \frac{1}{2}\frac{dP_2}{dg}$. Substituting (3.42) and (3.43) in (3.41) we obtain for the coefficients a_i, b_j, $0 \le i, j \le 4$ a linear algebraic system. Equalizing the coefficients in front of $f^m g^n$, $n \ge 0$, $m \ge 0$ we get:

$$a_0 = \mu, b_0 = \nu, b_2 = \lambda, \quad (3.44)$$

(λ, μ, ν are arbitrary real numbers),

$$a_1 = a_3 = b_1 = b_3 = 0, a_2 = 1 + \lambda, a_4 = -\nu, b_4 = -\mu.$$

Therefore, the sin-Gordon equation possesses the following solutions expressed by elliptic functions:

$$\varphi = 4\,arctg(f(x)g(t)), \quad (3.45)$$

where

$$(f')^2 = \mu f^4 + (1 + \lambda)f^2 - \nu, \quad (3.46)$$
$$(g')^2 = \nu g^4 + \lambda g^2 - \mu.$$

Example 4. a) Consider $\psi_1 = f_1(x)g_1(t)$, where $f_1(x) = c\,sh\frac{x}{\sqrt{1-c^2}}$, $c^2 < 1$, $g_1(t) = \frac{1}{ch\frac{ct}{\sqrt{1-c^2}}}$, $c \ne 0$. Then $\varphi_1 = 4\,arctg\psi_1$ satisfies (3.35).

b) $\psi_2 = f_2(x)g_2(t)$, $f_2(x) = sh\frac{ct}{\sqrt{1-c^2}}$, $0 < c^2 < 1$, $g_2(t) = \frac{1}{c\,ch\frac{x}{\sqrt{1-c^2}}}$. Then ψ_2 satisfies (3.35).

One can easily check that

$$\varphi = 4\,arctg\frac{tg\nu\,sin[(cos\,\nu)t]}{ch[(sin\,\nu)x]} \quad (3.47)$$

satisfies (3.35), $|\nu| < \frac{\pi}{2}$, ν being a parameter.

In physics the solution (3.47) is called a "breathon" ([76], Chapter 9).

Remark 3.2. It is worth mentioning that (3.35) is invariant under the Lorentzian transformations, i.e. if $\varphi(t, x)$ satisfies (3.35) then $\varphi(t', x')$ also satisfies (3.35) where (t', x') are given by the Lorentz change

$$\left| \begin{aligned} x' &= \frac{x - u_r t}{\sqrt{1 - u_r^2}}, \; |u_r| < 1 \\ t' &= \frac{t - u_r x}{\sqrt{1 - u_r^2}}, \; u_r = const. \end{aligned} \right.$$

Remark 3.3. In order to use formulas of the type (3.34) in studying (3.42), (3.43) the algebraic equations $P_1(f) = 0$, $P_2(g) = 0$ must have 4 real distinct roots. In the special case (3.46) the biquadratic equations must have two positive roots.

Evidently, the equation $az^2 + bz + c = 0$, $a \neq 0$ possesses two roots $z_1 > 0$, $z_2 > 0$ iff $\frac{c}{a} > 0$, $-\frac{b}{a} > 0$, $b^2 - 4ac > 0$. Under the same conditions the equation $aw^4 + bw^2 + c = 0$ possesses the roots $w_1 = \sqrt{z_1} > 0$, $w_2 = \sqrt{z_2}$ and $w_3 = -\sqrt{z_1}$, $w_4 = -\sqrt{z_2}$. This way we conclude that for $\nu > 0$ the equation (3.46) $P_1(f) = 0$ has 4 real distinct roots if $\mu < 0$, $\lambda > -1$, $(1 + \lambda)^2 > 4|\mu|\nu$, respectively $P_2(g) = 0$ has 4 real distinct roots iff $\mu < 0$, $\lambda < 0$, $\lambda^2 > 4|\mu|\nu$. Therefore, we can use various formulas for the elliptic integrals arising in (3.46) under the following conditions imposed on the parameters λ, μ, ν:

$$\nu > 0, \mu < 0, 2\sqrt{|\mu|\nu} < -\lambda < 1 - 2\sqrt{|\mu|\nu}, \sqrt{|\mu|\nu} < \frac{1}{4}. \tag{3.48}$$

Our next step is to discuss the interaction of the fluxons-antifluxons from physical point of view (see Chapter 9 from [76]). We shall assume that the velocity $c \in (0, 1)$ in Example 4a),b). To fix the ideas we shall deal with Example 4b). Having in mind that $sh \, x \approx \begin{cases} -\frac{1}{2}e^{-x}, & x \to -\infty \\ \frac{1}{2}e^x, & x \to \infty \end{cases}$ we shall find the asymptotic of $\psi_2 = \dfrac{sh \frac{ct}{\sqrt{1-c^2}}}{cch \frac{x}{\sqrt{1-c^2}}}$, respectively of $\varphi_2 = 4 \, arctg\psi_2$.

Put $ct = x + \xi\sqrt{1 - c^2}$, ξ-being fixed, i.e. $\frac{ct}{\sqrt{1-c^2}} = \frac{x}{\sqrt{1-c^2}} + \xi$, $|\xi| \leq \tilde{C} \Rightarrow$

$$\psi_2 \approx \begin{cases} -\frac{1}{c}e^{-\xi}, & x \to -\infty \\ \frac{1}{c}e^{\xi}, & x \to \infty \end{cases} = \frac{1}{c} \begin{cases} -e^{\frac{x-ct}{\sqrt{1-c^2}}}, & x \to -\infty, \; (IV) \\ e^{-\frac{x-ct}{\sqrt{1-c^2}}}, & x \to \infty, \; (II). \end{cases}$$

Both solutions are antifluxons as they are monotonically decreasing. We shall denote them by II and IV (see Fig. 3.7)

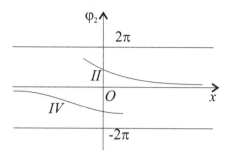

Fig. 3.7   (antifluxons)

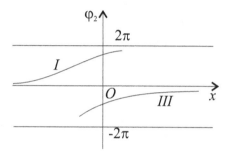

Fig. 3.8   (fluxons)

In a similar way we put $ct = -x + \eta\sqrt{1 - c^2}$, η - fixed and obtain:

$$\psi_2 \approx \frac{1}{c}\begin{cases} e^{\frac{x+ct}{\sqrt{1-c^2}}}, & x \to -\infty, \ (I) \\ -e^{-\frac{x+ct}{\sqrt{1-c^2}}}, & x \to \infty, \ (III) \end{cases}.$$

Both solutions are fluxons as they are monotonically increasing (see Fig. 3.8).

Consider (say) II. Evidently, $\frac{1}{c}e^{\frac{ct-x}{\sqrt{1-c^2}}} = e^{\frac{ct-x+\sqrt{1-c^2}ln\frac{1}{c}}{\sqrt{1-c^2}}}$, i.e. there is a shift of x: $x \to x - \sqrt{1 - c^2}ln\frac{1}{c}$.

The interaction fluxon-antifluxon is given in Fig. 3.9.

In Fig. 3.10 one can see the profile of the breathon for $\nu = \frac{\pi}{4}$.

We shall discuss briefly the interaction of a pair of two fluxons satisfying (3.35). The corresponding solution $\varphi(t, x)$ is given by the formula: $\varphi = 4\,arctg\frac{1-ABC}{B+C} = 4\,arctg\frac{\frac{1}{BC}-A}{\frac{1}{B}+\frac{1}{C}}$, where $A = \frac{1-c_1c_2-\sqrt{1-c_1^2}\sqrt{1-c_2^2}}{1+c_1c_2+\sqrt{1-c_1^2}\sqrt{1-c_2^2}}$,

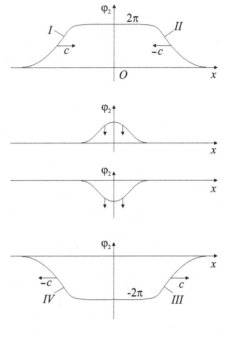

Fig. 3.9

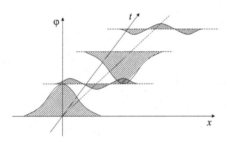

Fig. 3.10

$0 < c_1, c_2 < 1$, $B = e^{-\frac{x - x_1 - c_1 t}{\sqrt{1 - c_1^2}}}$, $C = e^{-\frac{x - x_2 - c_2 t}{\sqrt{1 - c_2^2}}}$. One can easily check that φ is a solution of (3.35) and that $0 \le A < 1$. Moreover, $A = 0 \iff c_1 = c_2$. Certainly, $c_2 > c_1 > 0$ (say) are called velocities of C, B, while x_1, respectively x_2 for $t = 0$ are the phases of the corresponding fluxons B, C. We are going to verify that $\varphi(t, x)$ describes for $t \to \pm\infty$ a pair of fluxons having the velocities c_1 and c_2. To do this we shall consider two cases:

1) Put $x - x_1 - c_1 t = \xi = const \Rightarrow B = e^{-\frac{\xi}{\sqrt{1-c_1^2}}}, C = e^{-\frac{x_1 - x_2 + t(c_1 - c_2) + \xi}{\sqrt{1-c_2^2}}}$.
Then

a) $t \to -\infty \Rightarrow C \to 0 \Rightarrow \varphi \to 4\,arctg\,e^{\frac{\xi}{\sqrt{1-c_1^2}}} = 4\,arctg\,e^{\frac{x - x_1 - c_1 t}{\sqrt{1-c_1^2}}}$ (I).
The last expression I describes a fluxon with phase x_1 and velocity c_1.

b) Let $t \to +\infty$. Then $C \to +\infty \Rightarrow \varphi \to -4\,arctg(Ae^{-\frac{\xi}{\sqrt{1-c_1^2}}})$. On the other hand, $arctg\,\alpha + arctg\frac{1}{\alpha} = \frac{\pi}{2}$ for each $\alpha > 0$. Therefore, $\varphi \to$
$-2\pi + 4\,arctg(\frac{1}{A}e^{\frac{x - x_1 - c_1 t}{\sqrt{1-c_1^2}}})$ (II).

Evidently, $\frac{1}{A}e^{\frac{x - x_1 - c_1 t}{\sqrt{1-c_1^2}}} = e^{\frac{x - y_1 - c_1 t}{\sqrt{1-c_1^2}}}$, where $y_1 = x_1 - \sqrt{1-c_1^2}ln\frac{1}{A} = x_1 + \sqrt{1-c_1^2}ln\,A$.

This way we have obtained for $t \to +\infty$ a fluxon with phase y_1 and velocity $c_1 > 0$.

Conclusion: After the collision with the second fluxon the first one gets the negative shift $y_1 - x_1 = \sqrt{1-c_1^2}ln\,A < 0$. Consequently, the slower fluxon moving to the right with velocity $c_1 > 0$ is shifting additionally backward.

In a similar way we investigate case 2):

2) $x - x_2 - c_2 t = \eta$, $\eta = const$.

Then a) $t \to -\infty \Rightarrow B \to +\infty \Rightarrow \varphi \to -2\pi + 4arctg(\frac{1}{A}e^{\frac{x - x_2 - c_2 t}{\sqrt{1-c_2^2}}})$ (III),
$\frac{1}{A}e^{\frac{x - x_2 - c_2 t}{\sqrt{1-c_2^2}}} = e^{\frac{x - y_2 - c_2 t}{\sqrt{1-c_2^2}}}$, where $y_2 = x_2 + \sqrt{1-c_2^2}ln\,A$.

Evidently,

b) $t \to \infty \Rightarrow B \to 0 \Rightarrow \varphi \to 4\,arctg\,e^{\frac{x - x_2 - c_2 t}{\sqrt{1-c_2^2}}}$ (IV).

This way we conclude that after the collision the second fluxon gets the positive shift $x_2 - y_2 = -\sqrt{1-c_2^2}ln\,A = \sqrt{1-c_2^2}ln\frac{1}{A} > 0$. Therefore, the faster fluxon moving to the right with velocity $c_2 > c_1 > 0$ is shifting additionally forward (see Fig. 3.11).

Some comments on Fig. 3.11. The slower (I) and the faster (III) fluxons are starting for $t = -\infty$, (III) being located to the left with respect to I and $I \subset \{\varphi > 0\}$, $III \subset \{\varphi < 0\}$. For $t = 0$ the fluxon I is joining III and they are forming the configuration V. Here $k = \dfrac{ln\,A + \dfrac{x_1}{\sqrt{1-c_1^2}} + \dfrac{x_2}{\sqrt{1-c_2^2}}}{\dfrac{1}{\sqrt{1-c_1^2}} + \dfrac{1}{\sqrt{1-c_2^2}}}$,

$\varphi(0, x) > 0$ for $x > k$ and $\varphi(0, x) < 0$ for $x < k$. Moreover, $\varphi(0, x) \to 2\pi$ for $x \to +\infty$ and $\varphi(0, x) \to -2\pi$ for $x \to -\infty$. For $t \gg 1$ the two fluxons are regenerating with the same velocities c_1, c_2 and the same profiles. They are

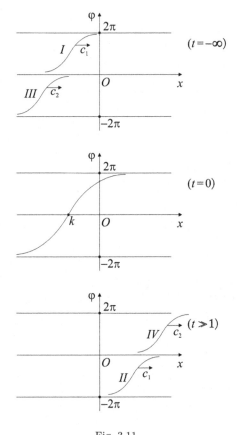

Fig. 3.11

denoted by IV (the faster one) and by II (the slower one). The fluxon IV is located to the right with respect to II, IV $\subset \{\varphi > 0\}$ and II $\subset \{\varphi < 0\}$.

Remark 3.4. Following J. Lamb [67] one can construct a series of solutions of (3.35) via the Backlund transformation. This transformation is described into details in [50], [92]. To do this we write (3.35) in the canonical form $\frac{\partial^2 \varphi}{\partial \xi \partial \eta} = sin\, \varphi$ after the change $\xi = \frac{x-t}{2}$, $\eta = \frac{x+t}{2}$.

The Backlund transformation is written as follows:

$$\varphi'_\xi = P(\varphi', \varphi, \varphi_\xi, \varphi_\eta, \xi, \eta)$$

$$\varphi'_\eta = Q(\varphi', \varphi, \varphi_\xi, \varphi_\eta, \xi, \eta), \frac{\partial}{\partial \eta}\varphi'_\xi = \frac{\partial}{\partial \xi}\varphi'_\eta$$

and in the special case of sin-Gordon equation it has the form

$$\frac{\partial \varphi'}{\partial \xi} = \frac{\partial \varphi}{\partial \xi} + 2\lambda \, sin \frac{\varphi' + \varphi}{2}, \lambda \neq 0$$

$$\frac{\partial \varphi'}{\partial \eta} = -\frac{\partial \varphi}{\partial \eta} + \frac{2}{\lambda} sin \frac{\varphi' - \varphi}{2}. \tag{3.49}$$

One can easily guess that $\frac{\partial}{\partial \eta} \varphi'_\xi = \frac{\partial}{\partial \xi} \varphi'_\eta$ implies that $\varphi_{\xi\eta} = sin \, \varphi$, while $\frac{\partial}{\partial \eta} \varphi'_\xi + \frac{\partial}{\partial \xi} \varphi'_\eta = 2\frac{\partial^2 \varphi'}{\partial \xi \partial \eta} = 2 \, sin \, \varphi'$.

Having in mind that $\varphi_0 = 0$ satisfies (3.35) we construct another solution φ_1 of (3.35) via the formula (3.49) with $\varphi = \varphi_0 = 0$. Thus, taking $\mu = \frac{1}{\lambda}$ we have:

$$\frac{\partial \varphi_1}{\partial \xi} = \frac{2}{\mu} sin \frac{\varphi_1}{2}, \frac{\partial \varphi_1}{\partial \eta} = 2\mu \, sin \frac{\varphi_1}{2}$$

$$\Rightarrow \frac{\partial \varphi_1}{\partial \eta} = \mu^2 \frac{\partial \varphi_1}{\partial \xi} \Rightarrow \varphi_1 = \gamma \left(\mu \eta + \frac{\xi}{\mu} \right), \gamma \in C^2.$$

To find $\gamma(z)$ we substitute the expression for φ_1 from the last equation in the second one and obtain:

$$\gamma'(z) = 2 \, sin \frac{\gamma}{2} \Rightarrow z = ln \left(\frac{1}{C} tg \frac{\gamma}{4} \right)$$

$$\Rightarrow tg \frac{\gamma}{4} = Ce^z = Ce^{\mu \eta + \frac{\xi}{\mu}}.$$

Taking $\mu = \sqrt{\frac{1-c}{1+c}}$ we get: $\varphi_1 = 4 \, arctg Ce^{\frac{x - ct}{\sqrt{1-c^2}}}, |c| < 1$.

Consider (3.49) and put $\varphi = \varphi_1$. Then for $\varphi_2 = \varphi'$ we obtain Example 4a) etc.

The Backlund approach can give a series of solutions of (3.35) eventually written in an explicit form. From the other hand, in many manuals on PDE one can find the following rather complete and elementary concerning the proof result on the solvability of the Cauchy problem for (3.35).

Theorem 3.3. *Consider the semilinear PDE*

$$\frac{\partial^2 u}{\partial \xi \partial \eta} = f(\xi, \eta, u, u_\xi, u_\eta) \tag{3.50}$$

in the rectangle $Q : \xi_0 \leq \xi \leq \xi_1, \eta_0 \leq \eta \leq \eta_1$. Suppose that the smooth curve $l : \eta = \eta(\xi), \eta' \neq 0$ connects the points $D = (\xi_0, \eta_1), B = (\xi_1, \eta_0)$ and *is contained in Q. Assume that f is continuous in $Q \times \mathbf{R}^3_{u,p,q}$ and satisfies the Lipschitz condition with respect to (u,p,q), i.e.*

$$|f(\xi,\eta,u_1,p_1,q_1) - f(\xi,\eta,u_2,p_2,q_2)| \le K(|u_1 - u_2| + |p_1 - p_2| + |q_1 - q_2|)$$

for some $K > 0$ and for each (ξ,η,u_1,p_1,q_1), $(\xi,\eta,u_2,p_2,q_2) \in Q \times \mathbf{R}^3$.
Then the Cauchy problem (3.50),

$$u|_l = \varphi_0(\xi) \in C^1[\xi_0,\xi_1] \qquad (3.51)$$
$$\frac{\partial u}{\partial \eta}|_l = \varphi_1(\xi) \in C^1[\xi_0,\xi_1]$$

possesses a unique classical solution in Q.

The conditions of Theorem 3.3 are satisfied by (3.35) as $|sin\,\varphi_1 - sin\,\varphi_2| \le |\varphi_1 - \varphi_2|$, $\forall \varphi_1, \varphi_2$.

The proof is based on the fixed point theorem.

In the last section of the chapter and for the sake of completeness we propose a detailed proof for the existence of a unique classical solution of the Cauchy problem to the sin-Gordon equation (3.35) with a right hand side $f(u)$ belonging to the Lipschitz class $C^{0,1}(\mathbf{R})$, i.e. $f \in C^0(\mathbf{R})$ and $|f(x) - f(y)| \le [f]_{C^{0,1}}|x - y|$, $\forall x,y \in \mathbf{R}$, $[f]_{C^{0,1}}$ being a constant.

4. We are going now to apply the results on elliptic functions formulated above to the travelling wave solutions of the CH-γ equation (3.2), respectively (3.7). This way we will obtain explicit formulas up to the inversion mapping theorem for the corresponding solutions. To begin with consider the polynomial of φ participating in the right-hand side of (3.7): $-P(\varphi) = \varphi^3 + A\varphi^2 + B\varphi + C$, where $A = c_0 - c$, $B = C_1$, $C = C_2$. Suppose that $\varphi_0 \in \mathbf{R}^1$ satisfies the algebraic cubic equation $P(\varphi_0) = 0$. Therefore, $-P(\varphi) = (\varphi - \varphi_0)(\varphi - p)(\varphi - q)$, i.e. according to the Viete formulas $p + q = -(A + \varphi_0)$, $B = pq - \varphi_0(A + \varphi_0)$, $-C = pq\varphi_0$. So p, q are the solutions of the equation $z^2 + (A + \varphi_0)z + B + (A + \varphi_0)\varphi_0 = 0$ and
$$p = \frac{-(A+\varphi_0)-\sqrt{A^2-4B-3\varphi_0^2}}{2}, q = \frac{-(A+\varphi_0)+\sqrt{A^2-4B-3\varphi_0^2}}{2}.$$
This is our main assumption concerning CH-γ equation.

(C) $\begin{cases} A^2 - 4B - 3\varphi_0^2 > 0 \iff (c - c_0)^2 - 4C_1 - 3\varphi_0^2 > 0, P(\varphi_0) = 0. \\ \text{The real numbers } p,q,\varphi_0,\gamma_0 = 1 + \frac{\gamma}{\alpha^2} \text{ are distinct.} \end{cases}$

Certainly, (C) implies that $p < q$ and

$$\alpha^2(\varphi')^2(\varphi - \gamma_0) = P(\varphi) = (\varphi - \varphi_0)(\varphi - p)(\varphi - q), \varphi(0) = \gamma_0. \qquad (3.52)$$

Example 5. Suppose that $\alpha^2 < 0$, (C) holds and $p < q < \gamma_0 < \varphi_0$, while $\varphi_0 \geq \varphi > \gamma_0$. According to the Theorem 3.1 case 1a), 3f) we have the configuration cusp-horizontal tangent. On the other hand, (3.52) gives us

$$H_1(\varphi) = \int_{\gamma_0}^{\varphi} \sqrt{\frac{\lambda - \gamma_0}{(\varphi_0 - \lambda)(\lambda - p)(\lambda - q)}}\, d\lambda = \frac{\xi}{\sqrt{-\alpha^2}}, 0 \leq \xi \leq \frac{T}{2}, \quad (3.53)$$

$$0 < \frac{T}{2} = \sqrt{-\alpha^2} \int_{\gamma_0}^{\varphi_0} \sqrt{\frac{\lambda - \gamma_0}{(\varphi_0 - \lambda)(\lambda - p)(\lambda - q)}}\, d\lambda < \infty.$$

The solution we are looking for is a periodic function with period T and such that $H_1(\varphi) = \frac{|\xi|}{\sqrt{-\alpha^2}}, |\xi| \leq \frac{T}{2}$.

Having in mind the formula 3.167, 22. from [100] we get that

$$H_1(\varphi) = \frac{2(\gamma_0 - q)}{\sqrt{(\varphi_0 - q)(\gamma_0 - p)}} \left[\Pi\left(\mu, \frac{\varphi_0 - \gamma_0}{\varphi_0 - q}, r\right) - F(\mu, r) \right], \quad (3.54)$$

where $\mu = arcsin\sqrt{\frac{(\varphi_0 - q)(\varphi - \gamma_0)}{(\varphi_0 - \gamma_0)(\varphi - q)}}, r = \sqrt{\frac{(\varphi_0 - \gamma_0)(q - p)}{(\varphi_0 - q)(\gamma_0 - p)}} < 1$.

Example 6. Assume that $\alpha^2 > 0$, (C) holds and $\gamma_0 < \varphi_0 < p < q$. In this case the Cauchy problem for (3.52) with initial data $\varphi(0) = \varphi_0$ possesses a classical periodic solution $\varphi = \varphi(\xi), p \geq \varphi > \varphi_0$ of the type 1c), 3f) (see Theorem 3.1).

Certainly,

$$H_2(\varphi) = \int_{\varphi_0}^{\varphi} \sqrt{\frac{\lambda - \gamma_0}{(\lambda - \varphi_0)(\lambda - p)(\lambda - q)}}\, d\lambda = \frac{|\xi|}{|\alpha|}, |\xi| \leq \frac{T}{2} \quad (3.55)$$

is the corresponding solution up to the inversion mapping theorem.

Applying formula 3.164, 4. from [100] we obtain that

$$H_2(\varphi) = \frac{2(\varphi_0 - \gamma_0)}{\sqrt{(q - \varphi_0)(p - \gamma_0)}} \Pi\left(\delta, \frac{p - \varphi_0}{p - \gamma_0}, s\right), \quad (3.56)$$

where $\delta = arcsin\sqrt{\frac{(p - \gamma_0)(\varphi - \varphi_0)}{(p - \varphi_0)(\varphi - \gamma_0)}}, s = \sqrt{\frac{(p - \varphi_0)(q - \gamma_0)}{(q - \varphi_0)(p - \gamma_0)}}$.

From our point of view it is preferable to work with the integral representation of the solution φ given by (3.53), (3.55) instead of the representation of φ by the elliptic integrals of first and third kind (3.54), (3.56). A glance on the integrals (3.53), (3.55) gives us immediately the qualitative structure of the solutions, while the investigation of (3.54), (3.56) seems to be more complicated. More specially, the full analysis of $\Pi(\varphi, n, k)$ is not trivial. We point out that the elliptic functions (3.29)-(3.31) give expressions only of the integrals having the form $\int R(x, \sqrt{P(x)})dx$, where R is a rational function and P is a polynomial of order 3 or 4 ([48], [2]). Therefore, even for polynomials of order ≥ 5 in the generic case one can not avoid the detailed study of integrals written as $\int \sqrt{\frac{Q(\lambda)}{P(\lambda)}}d\lambda$.

Remark 3.5. Consider again the equation (3.19). It was shown in [64] (see also [89]) that it possesses the following cuspon type solution with cusp at $\xi = 0(w = 1)$, $u(0) = -\frac{2}{3}k^2 < 0$:

$$u(\xi) = -\frac{32k^2}{3}\frac{w^2}{(3w^2+1)(w^2+3)}, \tag{3.57}$$

where

$$\xi = x + \frac{2k^2}{3}t + x_0, w = e^\xi + 2e^{\frac{\xi}{2}}\left[\left(sh\frac{\xi}{2}ch^2\frac{\xi}{2}\right)^{1/3} + \left(sh^2\frac{\xi}{2}ch\frac{\xi}{2}\right)^{1/3}\right].$$

It is proved also that $u(-\xi) = u(\xi)$.

An interesting question is to find this solution (3.57) or other travelling wave solutions of (3.19) in an explicit form by solving in elliptic functions the ODE (3.20) for appropriate C_1, C_2 ($C_2 = 0$) and then by "inverting" the corresponding equation of the form $H(\varphi) = |\xi|$, $H(\varphi)$ being expressed by (3.29), (3.30), (3.31). We point out that if $c = +\frac{2k^2}{3}$, $C_1 = C_2 = 0$ then (3.20) takes the form

$$\varphi^2\left(\varphi + \frac{8}{3}k^2\right) = (\varphi')^2\left(\varphi + \frac{2k^2}{3}\right), \varphi(0) = -\frac{2}{3}k^2, k \neq 0$$

and according to the Proposition 3.1 the solution φ satisfies an algebraic equation of order 6 with coefficients depending linearly on $e^{2\xi}$, $e^{4\xi}$. In fact, $\sqrt{\frac{C}{A}} = \frac{1}{2}$, i.e. $p = 1$, $q = 2$.

3.4   Cellular Neural Networks realization

Cellular Neural Networks (CNNs) are complex nonlinear dynamical systems, and therefore one can expect interesting phenomena like bifurcations and chaos to occur in such nets. It was shown that as the cell self-feedback coefficients are changed to a critical value, a CNN with opposite-sign template may change from stable to unstable. Namely speaking, this phenomenon arises as the loss of stability and the birth of a limit cycles.

Let us consider the modified sin-Gordon equation (3.35) in the following form:

$$\frac{\partial^2 \varphi}{\partial x^2} - \frac{\partial^2 \varphi}{\partial t^2} - \alpha(1 + \varepsilon \cos\varphi)\frac{\partial \varphi}{\partial t} = \sin\varphi. \tag{3.58}$$

This equation has the following physical meaning: it is the generalized version of the sin-Gordon equation (3.35) in Josephson junction [115]. In fact this equation arises in the well known Josephson Transmission Line (JTL) which is used in many applications in superconductor electronics. The relation between Cellular Neural Network (CNN) and the JTL array has been studied in many publications see for example [51], [115]. Two-dimensional array of Josephson Junctions is considered in [51]. Authors report the results of a Floquet analysis of such arrays of resistively and capacitively shunted JTLs in an external transverse magnetic field. The Floquet analysis indicates stable phase locking of the active junctions over a finite range of values of the bias current and junction capacitance, even in the absence of an external load.

For solving such an equation spatial discretization will be applied. The equation is transformed into a system of ordinary differential equations which is identified as the state equations of a CNN with appropriate templates. We map $\varphi(x,t)$ into a CNN layer such that the state voltage of a CNN cell at a grid point is u_j. Let us consider one-dimensional CNN, where the CNN cells consist of a linear capacitor in parallel with a nonlinear inductor described by $i_j = f(\varphi_j) = \alpha(1 + \varepsilon\cos\varphi_j)u_j - \sin\varphi_j$ and where these cells are coupled to each other by linear inductors with inductance L. In terms of CNN circuit topology we can identified the following corresponding elements:

1. CNN cell dynamics:

$$\frac{du_j}{dt} = \frac{1}{C}[I_j - f(\varphi_j)], \tag{3.59}$$

$$\frac{d\varphi_j}{dt} = u_j, 1 \le j \le N; \tag{3.60}$$

2. CNN synaptic law:

$$I_j = i_{L_j} - i_{L_{j+1}} = \frac{1}{L}(\varphi_{j-1} - 2\varphi_j + \varphi_{j+1}), \tag{3.61}$$

where $\varphi_j(t) = \int_{-\infty}^{t} u_j(\tau)d\tau$ is the flux-linkage at node j. Observe that the synaptic law (3.61) is a discrete Laplacian $A = [1, -2, 1]$ of the flux linkage φ_j.

Let us write the dynamics of our CNN model (3.59), (3.60), (3.61) in the following form:

$$\frac{du_j}{dt} = (\varphi_{j-1} - 2\varphi_j + \varphi_{j+1}) \tag{3.62}$$

$$- \alpha(1 + \varepsilon\, cos\, \varphi_j)u_j - sin\, \varphi_j$$

$$\frac{d\varphi_j}{dt} = u_j, 1 \leq j \leq N.$$

We shall study the dynamics of the above model (3.62) by applying the describing function method. Applying the double Fourier transform:

$$F(s,z) = \sum_{k=-\infty}^{k=\infty} z^{-k} \int_{-\infty}^{\infty} f_k(t)exp(-st)dt,$$

to the CNN model (3.62) we obtain:

$$sU(s,z) = (z^{-1}V(s,z) - 2V(s,z) + zV(s,z)) \tag{3.63}$$

$$- \alpha(1 + \varepsilon cosV(s,z))U(s,z) - sin\, V(s,z)$$

$$sV(s,z) = U(s,z).$$

Without loss of generality we can denote $N(U,V) = -\alpha\varepsilon cosV(s,z)U(s,z) - sin\, V(s,z)$. In the double Fourier transform we suppose that $s = iw_0$, and $z = exp(i\Omega_0)$, where w_0 is a temporal frequency, Ω_0 is a spatial frequency.

According to the describing function method, $H(s,z) = \frac{1}{s^2+s\alpha-(z^{-1}-2z+z)}$ is the transform function, which can be presented in terms of w_0 and Ω_0, i.e. $H(s,z) = H_{\Omega_0}(w_0)$.

We are looking for possible periodic state solutions of system (3.63) of the form:

$$X_{\Omega_0}(w_0) = X_{m_0}\, sin(w_0 t + j\Omega_0), \tag{3.64}$$

where $X = (U,V)$. According to the describing function method we take the first harmonics, i.e. $j = 0 \Rightarrow$

$$X_{\Omega_0}(w_0) = X_{m_0}\, sin\, w_0 t,$$

On the other side if we substitute $s = iw_0$ and $z = exp(i\Omega_0)$ in the transfer function $H(s,z)$ we obtain:

$$H_{\Omega_0}(w_0) = \frac{-iw_0^2\alpha^2 - (2\, cos\, \Omega_0 - 2 + w_0^2)}{w_0^2\alpha^2 + (2\, cos\, \Omega_0 - 2 + w_0^2)^2}. \tag{3.65}$$

According to (3.65) the following constraints hold:

$$\Re(H_{\Omega_0}(w_0)) = \frac{U_{m_0}}{V_{m_0}} \tag{3.66}$$

$$\Im(H_{\Omega_0}(w_0)) = 0.$$

Because our CNN model (3.62) is a finite circular array of N cells we have finite set of spatial frequencies:

$$\Omega_0 = \frac{2\pi k}{N}, 0 \leq k \leq N - 1. \tag{3.67}$$

Thus, (3.65), (3.66) and (3.67) give us necessary set of equations for finding the unknowns U_{m_0}, V_{m_0}, ω_0, Ω_0. As we mentioned above we are looking for a periodic wave solution of (3.62), therefore U_{m_0} and V_{m_0} will determine approximate amplitudes of the waves, and $T_0 = 2\pi/\omega_0$ will determine the wave speed.

Based on the above considerations the following proposition hold:

Proposition 3.2. *CNN model (3.62) of circular array of N identical, inductively coupled Josephson Junctions has periodic state solution $u_j(t)$, $\varphi_j(t)$ with a finite set of spatial frequencies $\Omega_0 = 2\pi k/N$, $0 \leq k \leq N - 1$.*

Simulating our CNN model (3.62) we obtain the following figure for the fluxon and breathon solution:

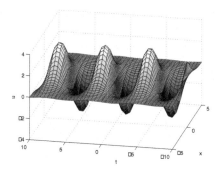

Fig. 3.12   Cellular wave computing based on the model (62).

Remark 3.6. By applying the describing function method we have been able to obtain a characterization of the periodic steady state solutions of our CNN model. In order to validate the accuracy of the achieved result it would be useful to have a possible initial condition from which the network will reach a steady state solution characterized by the desired value of Ω_0. One such possibility is to take a initial condition

$$u_j = \sin(\Omega_0 j), j = 1, 2, \ldots, N.$$

It should be mentioned that the continuous counterpart of the considered model is a modified sin-Gordon equation. A CNN model of the original sin-Gordon equation is presented in [109].

3.5   Cauchy problem for the sin-Gordon equation.

1. Following [105] we shall propose an interesting result concerning the global existence of the generalized solutions to the Cauchy problem

$$\Box u = \frac{\partial^2 u}{\partial t^2} - \frac{\partial^2 u}{\partial x^2} = f(u) \tag{3.68}$$

$$u|_{t=0} = u_0(x)$$

$$u_t|_{t=0} = u_1(x),$$

where $f \in C^{0,1}(\mathbf{R})$ and $(u_0, u_1) \in H^1_{loc} \times L^2_{loc}$.

Consequently, the right-hand side has a Lipschitz nonlinearity and the solution u is a weak one. Below H^s stands for the standard Sobolev class.

Definition 3.1. Let $f(u) \in L^1_{loc}(\mathbf{R}^1_t \times \mathbf{R}^1_x)$. We shall say that $u \in L^1_{loc}(\mathbf{R}^1_t \times \mathbf{R}^1_x)$ is a weak solution of the Cauchy problem (3.68) if for each $\varphi \in C^\infty_0(\mathbf{R}^1_t \times \mathbf{R}^1_x)$ the following identity holds:

$$\int_0^\infty \int_{\mathbf{R}_x} u \Box \varphi \, dz = \int_0^\infty \int_{\mathbf{R}_x} f(u)\varphi \, dz + \int_{\mathbf{R}_x} u_0 \partial_t \varphi(0,.) dx \tag{3.69}$$

$$- \int_{\mathbf{R}_x} u_1 \varphi(0,.) dx; \, z = (t,x).$$

Evidently, $L^2_{loc} \subset L^1_{loc}$. A weak finite energy solution is a weak solution $u \in L^1_{loc}(\mathbf{R}^1 \times \mathbf{R}^1)$ such that in addition $Du = (\partial_t u, \partial_x u) \in L^\infty_{loc}(\mathbf{R}^1_t : L^2_{loc}(\mathbf{R}^1_x))$. If the function u satisfies $Du \in C^0(\mathbf{R}^1 : L^2(\mathbf{R}^1))$ then u is called finite energy solution. We define solutions $u \in H^s$ for $s > 1$ if u is a finite energy solution and $Du \in C^0(\mathbf{R}^1 : H^{s-1}(\mathbf{R}^1))$.

Proposition 3.3 plays a crucial role in proving the global existence results to the Cauchy problem (3.68).

Proposition 3.3. *Consider the equation*

$$\Box u = \partial_t^2 u - \partial_x^2 u = h(t,x), t \ge 0$$

in the characteristic triangle $K^-(z_0) = \{z = (t,x) : |x - x_0| < t_0 - t\}$, $z_0 = (t_0, x_0)$, $t_0 > 0$ *with the boundary* $M^-(z_0) = \{|x - x_0| = t_0 - t\}$. *Put*

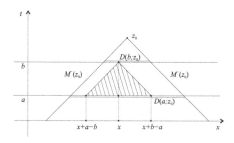

Fig. 3.13

$D(a; z_0) = K^-(z_0) \cap \{t = a\}$, $D(b; z_0) = K^-(z_0) \cap \{t = b\}$, $0 \leq a < b < t_0$. If u is finite energy solution and $h \in L^1_{loc}(\mathbf{R}^1 : L^2(\mathbf{R}^1_x))$ then

$$\|Du(b)\|_{L^2(D(b;z_0))} \leq \|Du(a)\|_{L^2(D(a;z_0))} \tag{3.70}$$

$$+ \|h\|_{L^1([a,b]:L^2(D(\tau;z_0)))}, a \leq \tau \leq b.$$

Proof. As the proof is simple we propose a short sketch only (for the sake of simplicity see Fig. 3.13).

According to D'Alembert's formula (combined with Duhamel's principle for non-homogeneous wave equations) we have that with initial data for $t = a$

$$u(t, x) = \frac{\tilde{u}_0(x + t - a) + \tilde{u}_0(x - t + a)}{2} \tag{3.71}$$

$$+ \frac{1}{2} \int_{x-t+a}^{x+t-a} \tilde{u}_1(y)dy + \frac{1}{2} \int_a^t \int_{x-t+\tau}^{x+t-\tau} h(\tau, y)dyd\tau;$$

u is a finite energy solution and by definition $\tilde{u}_0(x) = u(a, x)$, $\tilde{u}_1(x) = u_t(a, x) \Rightarrow \tilde{u}_0'(x) = \frac{\partial u}{\partial x}(a, x)$.

The characteristic sides of the exterior triangle are $\xi - x_0 = \tau - t_0$ and $\xi - x_0 = t_0 - \tau$, while the characteristic sides of the interior (depicted) triangle are $\xi - x = \tau - b$ and $\xi - x = b - \tau$, i.e. $a \leq t \leq b$, $x - b + \tau \leq \xi \leq x + b - \tau$. Differentiating (3.71) with respect to t, x and taking $t = b$ we get:

$$\frac{\partial u}{\partial x}\Big|_{t=b} = \frac{\tilde{u}_0'(x + b - a) + \tilde{u}_0'(x - b + a)}{2}$$

$$+ \frac{\tilde{u}_1(x + b - a) - \tilde{u}_1(x - b + a)}{2} + \frac{1}{2}\int_a^t [h(x + b - \tau) - h(x - b + \tau)]d\tau,$$

$$\frac{\partial u}{\partial t}\Big|_{t=b} = \frac{\tilde{u}_0'(x+b-a) - \tilde{u}_0'(x-b+a)}{2}$$

$$+\frac{\tilde{u}_1(x+b-a)+\tilde{u}(x-b+a)}{2} + \frac{1}{2}\int_a^t [h(\tau, x+b-\tau) + h(\tau, x-b+\tau)]d\tau.$$

Having in mind that $\| \int_a^t h(t,x)dt \|_{L^2(\mathbf{R}_x^1)} \leq \int_a^t \|h(\tau,x)\|_{L^2(\mathbf{R}_x^1)} d\tau$, $a \leq t \leq b$ we obtain (3.70). Then (3.70) implies that

$$\frac{\|Du(b)\|_{L^2(D(b;z_0))} - \|Du(a)\|_{L^2(D(a;z_0))}}{b-a} \tag{3.72}$$

$$\leq \frac{\|h\|_{L^1([a;b]:L^2(D(\tau,z_0)))}}{b-a}.$$

Fix $0 \leq a \leq z_0$ and let $b \to a$, $b > a$.
The inequality (3.72) implies

$$\frac{d}{da}\|Du(a)\|_{L^2(D(a))} \leq \|h\|_{L^2(D(a))} \tag{3.73}$$

for each $a \geq 0$.

Further on we shall work in the space of functions $u(t,x) \in C([a,b] : L^2(D(t,z_0)))$, $a \leq t \leq b$ (shortly, we write $C^0L^2(K^-)$).

2. We formulate below the following interesting result.

Theorem 3.4. *Let $f \in C^{0,1}(\mathbf{R}^1)$ and the initial data $(u_0, u_1) \in H^1_{loc} \times L^2_{loc}$. Then the Cauchy problem (3.68) possesses a unique global in time solution u and such that $Du \in C^0L^2_{loc}(\mathbf{R}^1 \times \mathbf{R}^1)$.*

Sketch of the proof. We shall construct u on a cone $K^-(z_0)$ for arbitrary $z_0 = (t_0, x_0)$. We can assume that $x_0 = 0$, $t_0 > 0$.

Put $X = \{u \in C^0([0, t_0] : L^2(D(t, z_0))), 0 \leq t \leq t_0$, such that $Du \in C^0([0, t_0] : L^2(D(t, z_0))), (u, u_t)|_{t=0} = (u_0, u_1)\}$. Suppose that $v(t, x) = L(u)$ solves the Cauchy problem

$$\Box v = f(u) = h(t,x) \tag{3.74}$$

$$v|_{t=0} = u_0$$

$$v_t|_{t=0} = u_1,$$

where $u \in X$ and we denote $h = f(u)$; $\Box(Lu) = f(u)$ and $u \in X \Rightarrow f(u) \in C^0L^2(K^-)$.

As we know the generalized solution v of (3.74) is given by D'Alembert's formula (3.71). If $u_0 \in C^2$, $u_1 \in C^1$ and $h \in C^1$ the only solution u is classical. It is important to point out that D'Alembert's formula still holds and gives a generalized solution u, $u \in L^2_{loc}$ for $u_0 \in H^1$, $u_1 \in L^2$ and $h \in L^2$. The mapping $L : X \to X$. According to the energy estimate (3.73)

$$\frac{d}{dt}||D[(Lu) - (Lw)]||_{L^2(D(t))} \tag{3.75}$$

$$\leq ||f(u) - f(w)||_{L^2(D(t))} \leq [f]_{C^{0,1}}||u - w||_{L^2(D(t))}$$

$$\leq [f]_{C^{0,1}} \int_0^t ||\partial_t(u - w)||_{L^2)(D(\tau))} d\tau$$

$$\leq [f]_{C^{0,1}} t_0 ||D(u - w)||_{L^\infty L^2(K^-)},$$

where

$$[f]_{C^{0,1}} = sup_{\lambda \neq \mu} \frac{|f(\lambda) - f(\mu)|}{|\lambda - \mu|},$$

$$||u||_{L^\infty L^2(K^-)} = ess\, sup_{0<t<t_0}||u(t,.)||_{L^2(D(t,z_0))}.$$

Certainly, $u(t,x) - u(0,x) = \int_0^t \partial_s u(s,x) dx \Rightarrow ||u(t,x) - u(0,x)||_{L^2(D(t))} \leq \int_0^t ||\partial_s u(s,x)||_{L^2(D(s))} ds$.

Integrating (3.75) we get:

$$||D[(Lu) - L(w)]||_{L^\infty L^2(K^-)} \leq [f]_{C^{0,1}} t_0^2 ||D(u - w)||_{L^\infty L^2(K^-)}, \tag{3.76}$$

because $D(Lu)|_{t=0} = D(Lw)|_{t=0} = (u_0', u_1)$.

The inequality (3.76) enables us to conclude that for $0 < t_0 < [f]_{C^{0,1}}^{-1}$ the nonlinear mapping L defines a contraction on the complete metric space X. According to the fixed point theorem there exists a unique fixed point $u \in X$ and therefore the Cauchy problem (3.68) admits a unique solution $u \in C^0 H^1(K^-(z_0))$. The main observation here is that $t_0 > 0$ depends only on $[f]_{C^{0,1}}$ and does not depend of the initial data. Fix some $t_0 > 0$ and consider the strip $[0, \frac{t_0}{2}] \times \mathbf{R}_x^1$. Evidently we may cover this strip with countably many cones of height $t_0 > 0$. Using the uniqueness in the overlapping domains we conclude that (3.68) possesses n unique solution u in $[0, \frac{t_0}{2}] \times \mathbf{R}_x^1$ and such that $Du \in C^0 L^2_{loc}([0, \frac{t_0}{2}] \times \mathbf{R}_x^1)$. Taking the initial data for $t = \frac{t_0}{2}$ we can construct (prolonge) the solution u in the strip $[\frac{t_0}{2}, t] \times \mathbf{R}_x^1$. Iterating the above described procedure we extend u for all $t \geq 0$.

This is our last observation.

Corollary 3.1. *Consider (3.68) with* $(u_0, u_1) \in H^2_{loc} \times H^1_{loc}$. *Then the solution* $u \in H^2_{loc}$ *and for any cone* $K^-(z_0) : D^2 u \in C^0 L^2(K^-)$.

The proof is trivial as

$$\Box(\partial_x u) = f'(u)\partial_x u, \ (\partial_x u, \partial_x u_t)|_{t=0} = (\partial_x u_0, \partial_x u_1) \in H^1_{loc} \times L^2_{loc}.$$

According to (3.73):

$$\frac{d}{dt}\|D\partial_x u\|_{L^2(D(t))} \leq \|f'(u)\partial_x u\|_{L^2(D(t))} \leq [f]_{C^{0,1}}\|\partial_x u(t)\|_{L^2(D(t))}.$$

We used above the well known facts that Lipschitz functions are almost everywhere differentiable and $|f'(u)| \leq [f]_{C^{0,1}}$, a.e. in $\mathbf{R}^1_u$.

Consequently, $\frac{d}{dt}\|D\partial_x u\|_{L^2(D(t))} \leq const$, $0 \leq t \leq t_0$ and therefore $\|D\partial_x u\|_{L^2(D(t))} \leq const$.

On the other hand, from (3.68) we have that $u_{tt} - u_{xx} = f(u) \Rightarrow u_{tt} \in C^0 L^2(K^-) \Rightarrow D^2 u \in C^0 L^2(K^-)$.

One can easily see that the more regular the initial data in Sobolev classes H^s and f in C^l, $l \geq 1$ are, the more regular the solution u of (3.68) in the corresponding Sobolev classes is.

Chapter 4

Stability of periodic travelling wave solutions for some classes of Korteweg-de Vries (KDV) type equations

4.1 Introduction and formulation of the main results

1. As we know one of the properties of dispersive nonlinear evolution equations is that usually they possess steadily translating waves - the so called travelling waves. By depending on specific boundary conditions on the wave's shape, for instance, in the case of water waves, these special states of motion can give rise to either solitary or periodic waves. Moreover, various nonlinear dispersive model equations are in sharp approximation to the governing equations for water waves. From these considerations, the problems about the stability of travelling waves and their existence as exact solutions of the dynamical equations are very important. The situation regarding periodic travelling waves is rather delicate. The stability and the existence of explicit formulas of these progressive wave trains have received little attention. A first study of these wavefronts is due to Benjamin in [14] with regard to the periodic steady solutions called cnoidal waves which were found initially by Korteweg and de Vries for the equation

$$u_t + uu_x + u_{xxx} = 0 \qquad (4.1)$$

Benjamin proposed an approach to the stability on cnoidal waves having the form

$$\varphi(\xi) = \beta_2 + (\beta_3 - \beta_2)cn^2 \left(\sqrt{\frac{\beta_3 - \beta_1}{12}} \xi, k \right)$$

$\beta_3 > \beta_1$. Angulo, Bona and Scialom gave in [7] a complete theory of stability of cnoidal waves for (4.1) with respect to perturbations of the same period. New theories of stability for the focusing nonlinear Schrödinger

69

equation and mKdV (modified KdV equation) have been obtained in [5]. It worth pointing out that in all these works the use of a sophisticated spectral theory for the periodic eigenvalue problem for the second order Lamé equation in Jacobi form was crucial.

2. It will be studied in this chapter the stability of the mKdV equation

$$u_t + u^2 u_x + u_{xxx} = 0, \tag{4.2}$$

i.e. $u_t + u^2 u_x - M u_x = 0$, $M = -\partial_x^2$.

The travelling wave solution of (4.2) has the form $u = \varphi_c(x - ct)$, $c > 0$, where the profile φ_c is a smooth periodic function with an a-priori fundamental period $L > 0$. Substituting $u = \varphi_c(x - ct)$ in (4.2) we obtain that $\varphi_c'' + \frac{1}{3}\varphi_c^3 - c\varphi_c = 0$, i.e.

$$(M + c)\varphi - \frac{1}{3}\varphi^3 = 0. \tag{4.3}$$

Certainly, we have integrated once (4.2), taking the integration constant 0.

Associated with (4.3) we consider the linear, closed, unbounded self adjoint operator $\mathcal{L} : D(\mathcal{L}) \to L_{per}^2([0, L])$ defined on a dense subspace of $L_{per}^2([0, L])$ by

$$\mathcal{L}u = (M + c)u - \varphi^2 u \tag{4.4}$$

From the theory of compact, symmetric operators applied to the periodic one dimensional eigenvalue problem

$$\left| \begin{array}{l} \mathcal{L}\psi = \lambda\psi \\ \psi(0) = \psi(L), \psi'(0) = \psi'(L) \end{array} \right. \tag{4.5}$$

it follows that the spectrum of $\mathcal{L}$ is a countable infinite set of eigenvalues $\{\lambda_n\}$ such that

$$\lambda_0 \leq \lambda_1 \leq \ldots \lambda_n \leq \ldots, \tag{4.6}$$

where $\lambda_n \to +\infty$.

It is evident that (4.3) implies that $\mathcal{L}$ has 0 as an eigenvalue with cigenfunction $\frac{d\varphi}{dx}$. In fact, $(M + c)\varphi' - \varphi^2\varphi' = 0$.

Below we propose the very important definition of stability of the orbit generated by φ_c, namely $\Omega_{\varphi_c} = \{\varphi_c(. + y) : y \in \mathbf{R}\}$. More precisely, if φ is L-periodic solution of (4.3), consider the translation $\tau_y\varphi(x) = \varphi(x + y)$, $x, y \in \mathbf{R}$. Then the orbit Ω_φ generated by φ is given as

$$\Omega_\varphi = \{g : g = \tau_y\varphi \text{ for some } y \in \mathbf{R}\},$$

$\Omega_\varphi \subset H_L^1$. For any $\eta > 0$ define the set $U_\eta \subset H_L^1$, $U_\eta = \{f : inf_{g \in \Omega_\varphi}||f - g||_{H_L^1} < \eta\}$.

Definition 4.1. The L-periodic solution of (4.3) is (orbitally) stable in H_L^1 by the flow generated by (4.2) if the following two conditions hold:

(i) There exists s_0 such that $H_L^{s_0} \subset H_L^1$ and the initial value problem for (4.2) is globally well posed in $H_L^{s_0}$.

(ii) For every $\varepsilon > 0$ there is $\delta > 0$ such that for all initial data $u_0 \in U_\delta \cap H_L^{s_0}$ the solution $u(t) = u(t,x)$ of (4.2) with $u(0,x) = u_0(x)$ is global in time and moreover $u(t) \in U_\varepsilon$ for each $t > 0$ fixed.

Concerning (i) we can use Theorem 1 proved in [27].

Theorem 4.1. *Let $s \geq 1$ be given. Then for each $u_0 \in H_L^s$ there exists a unique solution of (4.2) with $u|_{t=0} = u_0(x)$ and such that for every $T > 0$ $u(t)$ exists, $0 \leq t < T$ and $u \in C([0,T); H_L^s)$. Moreover, the mapping $u_0 \mapsto u$ is an analytic function between the corresponding function spaces.*

Thus, from the papers [13], [17], [53], [118] these are the conditions that imply stability:

(P_0) There exists a nontrivial smooth curve of real valued L-periodic solutions to (4.3) of the form $c \in I \subset \mathbf{R} \rightsquigarrow \varphi_c \in H^2[0,L]$, I being an interval,

(P_1) $\mathcal{L}$ has a unique negative eigenvalue λ and it is simple,

(P_2) the eigenvalue 0 of $\mathcal{L}$ is simple,

(P_3) $\frac{d}{dc} \int_0^L \varphi_c^2(x) dx > 0$.

The problem about the existence of a nontrivial smooth curve of periodic solutions of (4.2) having the same period $L > 0$ is a new and delicate aspect in proving of the orbital stability. In finding explicit solutions of (4.3) the theory of elliptic functions plays a fundamental role. Therefore, the solutions φ_c depend on the Jacobi elliptic functions of snoidal, cnoidal and dnoidal types as they are periodic with periods $4K(k)$, respectively $2K(k)$. This way we conclude that the elliptic modulus $k \in (0,1)$ will depend on the velocity c, i.e. the period of φ_c will possibly depend on c. By using the implicit function theorem one can construct the wanted smooth curve of periodic solutions having a fixed minimal period $L > 0$. As it concerns (P_3) it can be proved in some very special cases by combining the Poisson summation theorem and the Parseval's equality. The properties (P_1), (P_2) are obtained from the positivity of the Fourier transforms of some solitary waves satisfying (4.2). The analysis here proposed also relies upon the theory of totally positive operators and the class $PF(2)$ introduced by Karlin in [61] is basic in that study. The proof of (P_1), (P_2) is not trivial. We do not enter into details of the spectral theory and we do not verify that $(P_0) - (P_3)$ imply stability of the corresponding travelling waves as the spectral theory is out of the scope of our book.

Important note. The orbital stability results here proposed are obtained by periodic initial distributions **having the same minimal period** of our periodic solutions (L). Unfortunately, this approach does not work for obtaining stability results for more general periodic disturbances, for instance, for periodic disturbances $2L$.

This is the main theorem concerning orbital stability of (4.2). The proof relies heavily on the initial methods and techniques in [6] (see also [5]).

Theorem 4.2. [6] *Let φ_c be a periodic travelling wave solution of (4.3) and suppose that the part (i) of the definition of stability is verified. Assume also that the operator $\mathcal{L}$ (4.4) has the properties (P_1) and (P_2). Choose $\kappa \in L_L^2$ in such way that $\mathcal{L}\kappa = \varphi_c$ and put $I = (\kappa, \varphi_c)_{L_L^2}$. Then if $I < 0$, the periodic wave φ_c is orbitally stable.*

Remark 4.1. a) After a "bootstrap" arguments we can conclude that if $\varphi_c'' + \frac{1}{3}\varphi_c^3 - c\varphi_c = 0$, then $\varphi_c \in H_L^s$ for each $s \in \mathbf{R}^1$ and consequently $\varphi_c \in C^\infty[0, L]$, φ_c being periodic with period L.

b) Differentiating (4.3) with respect to c we get: $\mathcal{L}(\frac{\partial \varphi}{\partial c}) = -\varphi_c$, i.e. we can take $\kappa = -\frac{\partial \varphi}{\partial c}$ in Theorem 4.2 and we have to verify (P_0) and (P_3), as $I = -1/2 \frac{d}{dc} \|\varphi_c\|_{L_L^2}^2$.

For the sake of completeness we shall propose several definitions and several auxiliary results to be useful further on.

Let $\Omega \subset \mathbf{R}$ be an open interval on the real line and $1 \leq p \leq \infty$. Then the Lebesgue space $L^p(\Omega)$ is the standard set of measurable functions on Ω equipped with the norm

$$\|f\|_p = \left(\int_\Omega |f(x)|^p dx \right)^{1/p}.$$

In the special case $p = \infty$ we define $\|f\|_\infty = ess_\Omega \ sup|f(x)|$. The Banach space L^p is a Hilbert one for $p = 2$ and the scalar product in $L^2(\Omega)$ is given by $(f, g)_2 = \int_\Omega f(x)\overline{g(x)}dx$. These are the definitions of L_2 based on Sobolev spaces of periodic functions [15], [8]. Let $P = C_{per}^\infty$ be the set of all (complex valued) C^∞ periodic functions with period $l > 0$. Then P' stands for the set of all continuous linear functionals from P into $\mathbf{C}^1$ (i.e. the dual set of P). The elements of P' are called periodic distributions. For $k \in \mathbf{Z}$ put $\Theta_k(x) = e^{\frac{2\pi k i}{l} x}$, $x \in \mathbf{R}$. It is well known that $\{\Theta_k\}$ form a basis in P and in P' ([8]). The Fourier transform of $\psi \in P'$, namely $\hat{\psi} : \mathbf{Z} \to \mathbf{C}^1$, is defined as follows: $\hat{\psi}(k) = \frac{1}{l} < \psi, \Theta_{-k} >$, $\forall k \in \mathbf{Z}$. The numbers $\hat{\psi}(k)$ are called Fourier coefficients of the periodic function (distribution) ψ. As usual, $L_p[0,l] \hookrightarrow P'$ by defining $< \psi, \varphi >= \frac{1}{l} \int_0^l \psi(x)\varphi(x)dx$, $\forall \varphi \in P$ and $\psi \in L_p[0,l]$. Certainly, the Fourier coefficients $\hat{\psi}(k)$ of $\psi \in L_p[0,l]$ are given by $\hat{\psi}(k) = \frac{1}{l} \int_0^l \psi(x)e^{-\frac{2\pi kix}{l}} dx$.

Let $S(\mathbf{Z})$ be the space of rapidly decreasing sequences, i.e. sequences decreasing at infinity faster than any polynomial of $|n|$. Then the Fourier inverse transform of a sequence $\alpha = \{\alpha_k\}_{k \in \mathbf{Z}} \in S(\mathbf{Z})$ is the function $\alpha(x)$ defined by $\alpha(x) = \sum_{-\infty}^{\infty} \alpha_k \Theta_k(x)$. We denote by $C_{per,l}$ the space of the continuous and l-periodic functions. Consider a sequence $\alpha = \{\alpha_k\}_{k \in \mathbf{Z}}$ of complex numbers. Then the Hilbert space $l^2 = l^2(\mathbf{Z})$ is defined by

$$l^2 = \left\{ \alpha : ||\alpha||_{l^2} = \left(\sum_{-\infty}^{\infty} |\alpha_k|^2 \right)^{1/2} < \infty \right\}.$$

For arbitrary $s \in \mathbf{R}^1$ the Sobolev space $H_{per}^s[0,l] = H_l^s$ is the set of all $f \in P'$ such that

$$||f||_{H_l^s}^2 = l \sum_{-\infty}^{\infty} (1 + |k|^2)^s |\hat{f}(k)|^2 < \infty.$$

It is well known that H_l^s is a Hilbert space equipped with the scalar product

$$(f,g)_{H_l^s} = l \sum_{-\infty}^{\infty} (1 + |k|^2)^s \hat{f}(k)\overline{\hat{g}(k)}.$$

The Hilbert space H_l^0 is isometrically isomorphic to a subspace of $L^2([0,l])$ and $(f,g)_{H_l^0} = \int_0^l f(x)\overline{g(x)}dx$. The space H_l^0 will be denoted by L_l^2. Certainly, $H_l^s \subset L_l^2$ for all $s \geq 0$ and for $s > 1/2$ the Sobolev embedding theorem holds: $H_l^s \hookrightarrow C_{per,l}$. The Hilbert functional space $l_{s,l}^2 = l_{s,l}^2(\mathbf{Z})$ is defined by

$$l_{s,l}^2(\mathbf{Z}) = \left\{ \alpha = (\alpha_k)_{k \in \mathbf{Z}} : ||\alpha||_{l_{s,l}^2}^2 = l \sum_{-\infty}^{\infty} (1 + |k|^2)^s |\alpha_k|^2 < \infty \right\}.$$

We think that the notation l for the period, l^2, $l_{s,l}^2$ for the corresponding spaces do not lead to confusions.

One can easily see that $f \in H_l^s$ iff $\{\hat{f}(k)\}_{k \in \mathbf{Z}} \in l_{s,l}^2$ and from Parseval's theorem $||\hat{f}||_{l^2}^2 = \frac{1}{l}||f||_{L_l^2}^2$ it follows: $||f||_{H_l^s} = ||\hat{f}||_{l_{s,l}^2}$.

The convolution $\alpha * \beta$ of two sequences $\alpha = \{\alpha_k\}$, $\beta = \{\beta_k\}$ is a sequence defined by $(\alpha * \beta)_k = \sum_{j=-\infty}^{\infty} \alpha_{k-j}\beta_j$, $\forall k \in \mathbf{Z}$, whenever the series in the right-hand side are convergent $\forall k \in \mathbf{Z}$. Let $\alpha \in l^1(\mathbf{Z})$ and $\beta \in l^2(\mathbf{Z})$. Then according to the Young inequality $||\alpha * \beta||_{l^2} \leq ||\alpha||_{l^1} ||\beta||_{l^2}$, i.e. for each $\alpha \in l^1$ fixed the linear operator $\beta \in l^2 \mapsto \alpha * \beta \in l^2$ is continuous.

Let f, g be l-periodic sufficiently smooth functions. Then $f(x) = \sum_{-\infty}^{\infty} \hat{f}(k) e^{\frac{2\pi i k x}{l}}$, $\hat{f}(k) = \frac{1}{l} < f, e^{-\frac{2\pi i k x}{l}} >$, $g(x) = \sum_{-\infty}^{\infty} \hat{g}(k) e^{\frac{2\pi i k x}{l}}$, $\hat{g}(k) = \frac{1}{l} < g, e^{-\frac{2\pi i k x}{l}} >$.

Certainly, $\alpha = \{\alpha_k\} = \{\hat{f}(k)\}$, $\beta = \{\beta_k\} = \{\hat{g}(k)\}$ are the Fourier coefficients of f, respectively g and $\alpha, \beta \in l^2 \cap l^1$. Put $h(x) = f(x)g(x)$. Then $h(x) = \sum_{-\infty}^{\infty} \hat{h}(k) e^{\frac{2\pi i k x}{l}}$, where $\gamma_k = \hat{h}(k) = \frac{1}{l} \int_0^l h(x) e^{-\frac{2\pi i k x}{l}} dx = \frac{1}{l} \int_0^l f(x) \sum_{-\infty}^{\infty} \hat{g}(j) e^{-\frac{2\pi i x(k-j)}{l}} dx = \sum_{j=-\infty}^{\infty} \hat{g}(j) \hat{f}(k-j) = (\alpha * \beta)_k$.

This way we conclude that: $\alpha * \beta = \beta * \alpha$ and $\gamma = \{\gamma_k\} = \{(\alpha * \beta)_k\} = \alpha * \beta$, i.e. $\{\widehat{fg(k)}\} = \{\hat{f}(k)\} * \{\hat{g}(k)\}$. (Evidently, $\widehat{f^2} = \hat{f} * \hat{f}$). Thus, $\alpha = \{\alpha_k\}$, $\alpha_k > 0$, $\forall k \in \mathbf{Z}$ implies that $(\alpha * \alpha)_k > 0$, $\forall k \in \mathbf{Z}$.

Our last step is to introduce the class of sequences $PF(2)$ discrete.

Definition 4.2. The sequence $\alpha = \{\alpha_n\}_{n \in \mathbf{Z}} \subset \mathbf{R}^1$ belongs to the class $PF(2)$ discrete if:

(i) $\alpha_n > 0$, $\forall n \in \mathbf{Z}$

(ii) $\alpha_{n_1 - m_1} \alpha_{n_2 - m_2} - \alpha_{n_1 - m_2} \alpha_{n_2 - m_1} > 0$ for $n_1 < n_2$ and $m_1 < m_2$.

Definition 4.2 is a particular case of the Definition $PF(2)$ continuous which appears in [4] and [61]. Thus, the real-valued function $g(x)$, $x \in \mathbf{R}^1$ is in the class $PF(2)$ continuous if

(iii) $g(x) > 0$, $\forall x \in \mathbf{R}^1$

(iv) $g(x_1 - y_1)g(x_2 - y_2) - g(x_1 - y_2)g(x_2 - y_1) > 0$ for each $x_1, x_2, x_3, x_4 \in \mathbf{R}$ and such that $x_1 < x_2$ and $y_1 < y_2$.

It is shown in [3] that if $0 < g(x) \in C^2(\mathbf{R}^2)$, $\forall x \in \mathbf{R}^1$ and $g(x)$ is logarithmically concave, i.e. $\frac{d^2}{dx^2}(\log g(x)) < 0$ for $x \neq 0$ then $g \in PF(2)$ continuous.

As an illustrative example consider the function $g(x) = \mu sech(\nu x)$, where $\mu > o$, $\nu > 0$ are constants. Then $g \in PF(2)$ continuous and therefore $g \in PF(2)$ discrete as $\frac{d^2}{dx^2} \log g(x) = -\nu^2 sech^2(\nu x) < 0$, $\forall x$ $(sech\, x = \frac{1}{ch\, x})$.

We are formulating now an interesting result from [6] that guarantees the fulfilment of the conditions (P_1) and (P_2).

Theorem 4.3. [20] *Suppose that $\varphi_c > 0$ is even solution of (4.3) such that $\hat{\varphi}_c(n) > 0$, $\forall n \in \mathbf{Z}$ and $K = \widehat{\varphi_c^2} \in PF(2)$ discrete. Then both conditions (P_1), (P_2) hold for the operator $\mathcal{L}$ from (4.4).*

The proof is not elementary and certainly is based on several results from the spectral theory. Because of this reason and to simplify this book we omit it.

When investigating the orbital stability of the mKdV equation and without loss of generality we shall deal with (4.2) written in a more convenient form, namely: $u_t + 3u^2 u_x + u_{xxx} = 0$. Then (4.3) takes the form: $(M + c)\varphi - \varphi^3 = 0$ etc.

Concluding remarks. These are some comments on the validity of the conditions of the previous Theorem 4.3 (see [48]).

a) Consider the even continuous function $f(x) = x^2 + 1 > 0$, $x \in [-\pi, \pi]$ and periodic with period 2π on the real line. Then according to [48] $x^2 + 1 = \frac{\pi^2 + 3}{3} + 4 \sum_{n=1}^{\infty} (-1)^n \frac{\cos nx}{n}$, $\forall x \in [-\pi, \pi]$, i.e. its Fourier expansion possesses infinitely many positive and negative Fourier coefficients;

b) Let $g(x) = \frac{3x^2 - 6\pi x + 2\pi^2}{12}$ for $x \in [0, \pi]$. Then we continue $g(x)$ in an even way on $[-\pi, 0]$: $g(-x) = g(x)$ and then we continue it as 2π-periodic function on the real line. According to [48]: $g(x) = \sum_{n=1}^{\infty} \frac{\cos nx}{n^2}$, $x \in [0, \pi]$ and $g(0) > 0$, $g(\pi) < 0$, $g'(x) < 0$ for $x \in [0, \pi)$, $g'(\pi) = 0$. Therefore, the Fourier coefficients of $g(x)$ are positive but $g(x)$ changes its sign from " $+$ " to " $-$ " at the interior point $0 < \pi(1 - \frac{\sqrt{3}}{3}) < \pi$.

c) $\varphi_c > 0$ is a very important assumption for the validity of the conditions $\hat{\varphi}_c > 0$, $K = \widehat{\varphi_c^2} \in PF(2)$ of Theorem 4.3. In fact, assume that $\varphi_c(0) = 0$ and $\varphi_c \not\equiv 0$, $\varphi_c(-x) = \varphi_c(x)$. Then the Fourier expansion formula for φ_c ($0 \leq \varphi_c^2$) gives us: $\varphi_c(0) = \sum_{-\infty}^{\infty} \hat{\varphi}_c(n)$ ($\varphi_c^2(0) = \sum_{-\infty}^{\infty} \widehat{\varphi_c^2}(n)$). Thus there exists at least one pair of Fourier coefficients $\hat{\varphi}_c(n_1)$, $\hat{\varphi}_c(n_2)$ such that $\hat{\varphi}_c(n_1) . \hat{\varphi}_c(n_2) < 0$ ($\widehat{\varphi_c^2}(m_1) . \widehat{\varphi_c^2}(m_2) < 0$). Thus, $K \notin PF(2)$ discrete.

4.2 Some results from the theory of elliptic functions and proof of Theorem 4.2

1. We shall propose here several results useful in studying the stability of L-periodic travelling wave solutions for equations of KdV type and more specially the mKdV equation (4.2).

Consider the mKdV equation

$$u_t + 3u^2 u_x + u_{xxx} = 0. \tag{4.7}$$

Put $u = \varphi(x - ct) \equiv \varphi_c(\xi)$, where $\xi = x - ct$, $c > 0$ and suppose that φ_c is periodic function with period $L > 0$: $\varphi_c(\xi + L) = \varphi_c(\xi)$, $\forall \xi \in \mathbf{R}^1$.

Substituting $\varphi_c(\xi)$ in (4.7) we obtain

$$\varphi_c''' - c\varphi_c' + 3\varphi^2\varphi' = 0,$$

i.e. $\varphi_c'' + \varphi_c^3 - c\varphi_c = const.$ As we mentioned in Section 1 we will take $const = 0$. Evidently,

$$\varphi_c'' + \varphi_c^3 - c\varphi_c = 0 \tag{4.8}$$

and obviously,

$$(\varphi_c')^2 = d + c\varphi_c^2 - \frac{1}{2}\varphi_c^4, d = const. \tag{4.9}$$

Certainly, (4.9) is an ODE with separate variables and we can find its periodic solutions in integral form, expressing the corresponding integrals by the elliptic functions E, F,Π (see (3.29), (3.30), (3.31) from Chapter III). To do this we can use [48], [2], [77]. In order to simplify our problem and to avoid the use of more complicated functions we are looking for

$$\varphi_c(\xi) = \eta dn(\mu\xi, k) \tag{4.10}$$

$0 < dn(w,k)$ being defined as $h(w,k)$ from system (3.24) Chapter III. Here $\eta > 0$, $\mu > 0$ are constants.

Certainly, (4.10) is periodic with period $\frac{2K(k)}{\mu}$, i.e. we must take $L = \frac{2K(k)}{\mu}$. Substituting (4.10) in (4.8) and taking into account the formulas $(snw)' = cnwdnw$, $(cnw)' = -snwdnw$, $(dnw)' = -k^2snwcnw$, $sn^2w + cn^2w = 1$, $k^2sn^2w + dn^2w = 1$ and the definition $(k')^2 = 1 - k^2$ we get:

$$-2k^2\mu^2cn^2w + \mu^2k^2 + \eta^2((k')^2 + k^2cn^2w) - c = 0,$$

i.e.

$$\eta^2 = 2\mu^2 \iff \mu = \frac{\eta}{\sqrt{2}}, c = \mu^2k^2 + \eta^2(k')^2 \iff c = \frac{\eta^2}{2}(1 + (k')^2) \tag{4.11}$$

Consequently, (4.10) is an L-periodic solution of (4.8) iff $\varphi_c(\xi) = \eta dn(\frac{\eta\xi}{\sqrt{2}}, k)$, where $c = \frac{\eta^2}{2}(1 + (k')^2)$, $k' \in (0,1)$ and $\eta = \frac{2\sqrt{2}K(k)}{L}$ (as $L = \frac{2K(k)}{\mu}$). Evidently, $c = \frac{4K^2(k)(2-k^2)}{L^2}$, respectively the period $L = \frac{2K\sqrt{2-k^2}}{\sqrt{c}}$; $k^2 = 2 - \frac{2c}{\eta^2}$.

Remark 4.2. As we know $dn(w,k)$ is a periodic positive function having period $2K(k)$. Then according to [2] this is the Fourier expansion of the function $dn(\frac{2K\xi}{L}, k)$ having period L:

$$\frac{2K}{L}dn\left(\frac{2K\xi}{L}, k\right) = \frac{\pi}{L} + \frac{4\pi}{L}\sum_{n=1}^{\infty}\frac{q^n}{1+q^{2n}}cos\left(\frac{2n\pi\xi}{L}\right), \tag{4.12}$$

where $q = e^{-\pi K'/K}$, $\xi \in \mathbf{R}^1$.

2. In order to prove Theorem 4.2 we need some auxiliary results from the theory of the elliptic functions. Thus, remind the definition

$$0 < E(k) = \int_0^{\frac{\pi}{2}} \sqrt{1 - k^2 sin^2\alpha} d\alpha, k \in (0,1). \qquad (4.13)$$

As it is standard, we put $0 < K'(k) = K(k')$, $E'(k) = E(k')$. Then according to [48], [2], [77]:

$$\frac{dK}{dk} = \frac{E - (k')^2 K}{k(k')^2}, \frac{dE}{dk} = \frac{E - K}{k}, \qquad (4.14)$$

$K(0) = E(0) = \frac{\pi}{2}$, $E(1) = 1$, $K(1) = +\infty$.

Evidently, (3.26) implies that $E(k) < K(k)$, $\forall k \in (0,1)$ and $E(k) - (k')^2 K = k^2 \int_0^{\frac{\pi}{2}} \frac{cos^2\alpha}{\sqrt{1-k^2 sin^2\alpha}} d\alpha > 0$, i.e.

$$\frac{dK}{dk} > 0, \frac{dE}{dk} < 0 \quad \text{for} \quad k \in (0,1). \qquad (4.15)$$

It is easy to see that

$$\frac{dK'}{dk} = \frac{dK(k')}{dk'} \frac{(-k)}{k'} = -\frac{E' - k^2 K'}{k(k')^2} < 0, \forall k \in (0,1). \qquad (4.16)$$

So $k \in (0,1) \Rightarrow K(k) > \frac{\pi}{2}$.

This is another very important inequality to be useful further on. Put $P(k) = K^2(k)(2 - k^2) = K^2(k)(1 + (k')^2)$, $\forall k \in (0,1)$. Then

$$\frac{dP(k)}{dk} > 0. \qquad (4.17)$$

In other words,

$$\frac{d}{dk}(K(k)\sqrt{2 - k^2}) > 0, \forall k \in (0,1). \qquad (4.18)$$

We shall verify now (4.17). Obviously, (4.14) implies: $\frac{dP}{dk} = \frac{2K}{k(k')^2} A(k)$, where $A(k) = (2 - k^2)E - 2(1 - k^2)K = 2(E - K) - k^2(E - 2K)$. Therefore, we must prove that $A(k) > 0$, $\forall k \in (0,1)$. From [48] we know that

$$E(k) = \frac{\pi}{2} \left\{ 1 - \sum_{n=1}^{\infty} \left[\frac{(2n-1)!!}{(2n)!!} \right]^2 \frac{k^{2n}}{2n-1} \right\},$$

$$K(k) = \frac{\pi}{2} \left\{ 1 + \sum_{n=1}^{\infty} \left[\frac{(2n-1)!!}{(2n)!!} \right]^2 k^{2n} \right\}.$$

Consequently,

$$2(E - K) = -\pi \sum_{n=1}^{\infty} \left[\frac{(2n-1)!!}{(2n)!!} \right]^2 k^{2n} \frac{2n}{2n-1} \qquad (4.19)$$

$$-k^2(E - 2K) = \frac{k^2\pi}{2} + \pi k^2 \sum_{n=1}^{\infty} \left[\frac{(2n-1)!!}{(2n)!!} \right]^2 k^{2n} \frac{4n-1}{4n-2}. \qquad (4.20)$$

Combining (4.19), (4.20) we get:

$$A(k) = -\pi \sum_{n=2}^{\infty} \left[\frac{(2n-1)!!}{(2n)!!} \right]^2 k^{2n} \frac{2n}{2n-1} + \pi \sum_{n=1}^{\infty} \left[\frac{(2n-1)!!}{(2n)!!} \right]^2 k^{2n+2} \frac{4n-1}{4n-2}.$$

After the change $n + 1 \to n$, $n \geq 1$ in the second sum of the expression for $A(k)$ we have:

$$A(k) = \pi \left\{ -\sum_{n=2}^{\infty} \left[\frac{(2n-1)!!}{(2n)!!} \right]^2 k^{2n} \frac{2n}{2n-1} + \sum_{n=2}^{\infty} \left[\frac{(2n-3)!!}{(2n-2)!!} \right]^2 k^{2n} \frac{4n-5}{4n-6} \right\}$$

$$= \pi \sum_{n=2}^{\infty} \left[\frac{(2n-3)!!}{(2n-2)!!} \right]^2 k^{2n} \left(\frac{4n-5}{4n-6} - \frac{2n-1}{2n} \right) > 0$$

as $4n - 5 > 4n - 6 > 0$, $0 < \frac{2n-1}{2n} < 1$.

The proof of (4.17) is completed. This way we conclude that if $0 < k < 1$ then

$$P(k) > 2K^2(0) = \frac{\pi^2}{2}. \qquad (4.21)$$

There is another simpler way in proving that $A(k) > 0$. In fact, $\frac{dA(k)}{dk} = -2kE + (2-k^2)\frac{dE}{dk} + 4kK - 2(1-k^2)\frac{dK}{dk}$. Applying the formulas (4.14) it is easy to see that $\frac{dA}{dk} = 3k(K - E) > 0$, i.e. $A(k)$ is strictly monotonically increasing function. Then $A(0) = 0$ implies that $A(k) > 0$ for $k \in (0, 1)$.

We shall now remind the reader of the following famous Poisson summation theorem (see [110], [111]).

Theorem 4.4. *Let f, $\hat{f} \in C^0(\mathbf{R}^1)$, where $\hat{f}(x) = \int_{-\infty}^{\infty} f(y)e^{-ixy}dy$ and the inverse Fourier transform formula holds: $f(y) = \frac{1}{2\pi} \int_{-\infty}^{\infty} \hat{f}(x)e^{ixy}dx$. Moreover, we assume that the following inequalities are satisfied with some $\delta > 0$, $A > 0$, $\delta, A = const$: $|f(y)| \leq \frac{A}{(1+|y|)^{1+\delta}}$, $|\hat{f}(x)| \leq \frac{A}{(1+|x|)^{1+\delta}}$. Then for each $L > 0$ we have*

$$\sum_{n=-\infty}^{\infty} f(x + Ln) = \frac{1}{L} \sum_{n=-\infty}^{\infty} \hat{f}\left(\frac{n}{L}\right) e^{\frac{2i\pi nx}{L}}. \qquad (4.22)$$

Both series in (4.22) converge absolutely and $g(x + L) = g(x)$, $\forall x$, i.e. $g(x) = \sum_{n=-\infty}^{\infty} f(x + Ln)$ is continuous periodic function.

Fix the constant $\omega > 0$ and consider the mKdV equation with parameter $\omega > 0$:

$$\varphi_\omega'' + \varphi_\omega^3 - \omega\varphi_\omega = 0. \qquad (4.23)$$

One can easily guess that the function $\varphi(x) = a\,sech(bx)$, $a > 0$, $b > 0$, satisfies (4.23) if and only if $a = \sqrt{2}b$, $b = \sqrt{\omega}$, i.e.

$$\Phi_\omega(x) = \sqrt{2\omega}\,sech(\sqrt{\omega}x),\ \Phi_\omega(-x) = \Phi_\omega(x), \forall x \in \mathbf{R}^1$$

is a positive soliton type solution of (4.23).

As it is known [77], [87] the Fourier transform $\hat{\Phi}_\omega$ of Φ_ω is

$$\hat{\Phi}_\omega(x) = \sqrt{2\pi}\,sech\left(\frac{\pi x}{2\sqrt{\omega}}\right) \tag{4.24}$$

Applying (4.22) and using the fact that $sech(x)$ is even function we obtain that the L-periodic continuous function $\Psi_\omega(\xi) = \sum_{n=-\infty}^{\infty} \Phi_\omega(\xi+Ln)$ can be written in the following form:

$$\Psi_\omega(\xi) = \frac{\sqrt{2\pi}}{L} \sum_{n=-\infty}^{\infty} sech\left(\frac{\pi n}{2L\sqrt{\omega}}\right) e^{\frac{2\pi i n\xi}{L}}$$

$$= \frac{\sqrt{2\pi}}{L} \sum_{n=0}^{\infty} \varepsilon_n\, sech\left(\frac{\pi n}{2L\sqrt{\omega}}\right) \cos\left(\frac{2\pi n\xi}{L}\right), \tag{4.25}$$

where $\varepsilon_0 = 1$, $\varepsilon_n = 2$ for $n > 0$.

Certainly, $\Psi_\omega(\xi)$ is in some sense a periodisation of the soliton solution $\Phi_\omega(x)$ of (4.23).

We shall summarize now the main results given by (4.10), (4.12) and (4.25). Thus we rewrite (4.12) with $q = e^{-\pi K'/K}$ in the form:

$$\frac{2K}{L}dn\left(\frac{2K\xi}{L}, k\right) = \frac{\pi}{L} + \frac{2\pi}{L} \sum_{n=1}^{\infty} sech\left(\frac{n\pi K'}{K}\right) \cos\left(\frac{2\pi n\xi}{L}\right), \tag{4.26}$$

while $\varphi_c(\xi) = \eta dn(\frac{n\xi}{\sqrt{2}}, k)$ is an L-periodic solution of (4.8) iff

$$c = \frac{\eta^2}{2}(2 - k^2), k \in (0,1) \ \text{ and } \ \eta = 2\sqrt{2}\frac{K(k)}{L}. \tag{4.27}$$

So $k \in (0,1) \Rightarrow \eta \in (\sqrt{c}, \sqrt{2c})$ and $c = \frac{4K^2(k)(2-k^2)}{L^2} > \frac{2\pi^2}{L^2}$ since according to (4.17) the function $4K^2(k)(2 - k^2)$ is strictly monotonically increasing.

Define now

$$\omega = \frac{c}{16(2 - k^2)(K')^2(k)} = \frac{K^2(k)}{4L^2(K')^2(k)} \tag{4.28}$$

with $k \in (0,1)$.

Then it is easy to check that $\Psi_{\omega(c)} = \varphi_c$ is an L-periodic solution of (4.8).

Proposition 4.1. (P_0). *There exists a nontrivial smooth curve $c \in I \subset \mathbf{R}^1 \to \varphi_c \in H^n_{per}[0,L]$, $n \in \mathbf{N}$, of L-periodic (dnoidal type) wave solutions of (4.8), I being an interval on the real line and such that*

$$(P_3) \qquad \frac{d}{dc}\int_0^L \varphi_c^2(x)dx > 0.$$

Proof: (P_0). Put $T_{\varphi_c}(\eta) = \frac{2K(k(\eta))\sqrt{2-k^2(\eta)}}{\sqrt{c}}$, where $k^2(\eta) = 2 - \frac{2c}{\eta^2}$ (see (4.27)). Certainly, $T_{\varphi_c}(\eta)$ is the fundamental period of φ_c. According to (4.18) $T_{\varphi_c}(\eta) > \pi\sqrt{\frac{2}{c}}$. Therefore, the mapping $T_{\varphi_c} : \eta \in (\sqrt{c}, \sqrt{2c}) \to (\pi\sqrt{\frac{2}{c}}, +\infty)$ is a diffeomorphism as $T_{\varphi_c}(\sqrt{c}) = \pi\sqrt{\frac{2}{c}}$, $T_{\varphi_c}(\sqrt{2c}) = +\infty$.

Now we are ready to construct a family of dnoidal wave solutions having period $L > 0$. To do this we take $c > 0$ such that $c > \frac{2\pi^2}{L^2}$. Then $L > \pi\sqrt{\frac{2}{c}}$ implies that there is a unique solution $\eta(c) \in (\sqrt{c}, \sqrt{2c})$ of the equation $T_{\varphi_c}(\eta(c)) = L$. The function $\eta(c)$ depends smoothly on c. Put $V(k) = K(k)\sqrt{2-k^2}$. Then $T_{\varphi_c}(\eta(c)) = L \Rightarrow V(k(\eta(c))) = \sqrt{c}\frac{L}{2}$. Differentiating with respect to c we get: $\frac{dV}{dk}\cdot\frac{dk}{d\eta}\cdot\frac{d\eta}{dc} = \frac{1}{4\sqrt{c}L} > 0$. As according to (4.18) $\frac{dV}{dk} > 0$ we have: $\frac{dk}{dc} = \frac{dk}{d\eta}\cdot\frac{d\eta}{dc} > 0$. On the other hand, differentiating $k^2 = 2 - \frac{2c}{\eta^2}$ with respect to c we obtain: $2k\frac{dk}{dc} = -\frac{2\eta^2 - 4c\eta\frac{d\eta}{dc}}{\eta^4} \Rightarrow k\frac{dk}{dc} = \eta^{-3}(2c\frac{d\eta}{dc} - \eta) \Rightarrow 2c\frac{d\eta}{dc} - \eta > 0 \Rightarrow \frac{d\eta}{dc} > 0$.

This way we have shown that the mapping $c \in (\frac{2\pi^2}{L^2}, +\infty) \to \varphi_c \in H^n_{per}[0,L]$ is a smooth curve as the mapping $c \in (\frac{2\pi^2}{L^2}, \infty) \to \eta = \eta(c)$ is smooth and moreover, strictly monotonically increasing.

Technically, the following inequality is useful in proving (P_3): $\frac{d\omega}{dc} > 0$, where the mapping $\omega : (\frac{2\pi^2}{L^2}, +\infty) \to \mathbf{R}$ is defined by (4.28).

Indeed, $16\frac{d\omega}{dc} = \frac{(2-k^2)(K')^2 + 2c\frac{dk}{dc}K'(kK'-(2-k^2)\frac{dK'}{dk})}{(2-k^2)^2(K')^4}$.

Having in mind that according to (4.16) $\frac{dK'}{dk} < 0$ and that $\frac{dk}{dc} > 0$ we obtain $\frac{d\omega}{dc} > 0$.

Proof of (P_3): Consider $I = -\frac{1}{2}\frac{d}{dc}\int_0^L \varphi_c^2(\xi)d\xi$ and apply Parseval's equality: $I = -\frac{1}{2}\frac{d}{dc}\|\varphi_c\|^2_{L^2_{per}[0,L]} = -\frac{L}{2}\frac{d}{dc}\|\hat\varphi_c\|_{l^2}$. Then the identity $\varphi_c = \Psi_{\omega(c)}$ and (4.25) $(\hat\varphi_c(n) = \frac{\sqrt{2}\pi}{L}\varepsilon_n \, sech(\frac{\pi n}{2L\sqrt\omega}) > 0)$ give us:

$$\|\hat\varphi_c\|^2_{l^2} = \frac{2\pi^2}{L^2}\sum_{n=-\infty}^\infty sech^2\left(\frac{\pi n}{2L\sqrt\omega}\right)$$

$$\Rightarrow \frac{d}{dc}\|\hat\varphi_c\|^2_{l^2} = C_1(L)\frac{1}{\omega^{3/2}}\frac{d\omega}{dc}\sum_{n=-\infty}^\infty sech^2\left(\frac{\pi n}{2L\sqrt\omega}\right)nth\left(\frac{\pi n}{2L\sqrt\omega}\right),$$

with some constant $C_1(L) > 0$. The observation that $nth\frac{\pi n}{2L\sqrt{\omega}} = |n|th\frac{\pi|n|}{2L\sqrt{\omega}} > 0$ for $n \neq 0$, $th(0) = 0$ completes the proof (certainly, $\omega = \omega(c)$).

Now we are ready to prove the orbital stability of the periodic travelling solutions of (4.7). To do this we shall apply Theorems 4.3, 4.4 and we shall use the above made considerations. The only facts to be verified are: $\hat{\varphi}_c(n) > 0$ and $K = \widehat{\varphi_c^2}(n)$ belong to $PF(2)$ discrete. The identity $\varphi_c = \Psi_{\omega(c)}$ and $\varphi_c > 0$, $\varphi_c(-x) = \varphi_c(x)$, (4.25) give us that $\hat{\varphi}_c(n) = \frac{\sqrt{2\pi}}{L}\varepsilon_n \, sech(\frac{\pi n}{2L\sqrt{\omega}}) > 0$, $\hat{\varphi}_c(n) = \hat{\varphi}_c(-n)$. As we know, the function $g(x) = \mu \, sech(\nu x)$ belongs to the space $PF(2)$ continuous for $\nu > 0$, $\mu > 0$ - constants. Therefore, $\hat{\varphi}_c$ belongs to $PF(2)$ discrete. So, since according to [61] the convolution of even sequences in $PF(2)$ discrete is a sequence belonging to $PF(2)$ discrete we conclude that $K = \widehat{\varphi_c^2} \in PF(2)$ discrete. In fact, $\widehat{\varphi_c^2} = \hat{\varphi}_c * \hat{\varphi}_c$.

Conclusion. The dnoidal wave $\varphi_c = \Psi_{\omega(c)}$ with ω given by (4.28) is stable in $H^1_{per}([0, L])$ by the flow of the mKdV (4.7). This way Theorem 4.2 is proved.

The reader can enlarge his knowledge on the stability of solitary waves and periodic travelling waves by studying the papers proposed below in the references: [1], [54], [79], [82], [83], [84], [113], [74].

Footnote. We shall complete this Chapter with several results concerning the stability of the peakons for the Camassa-Holm equation

$$u_t - u_{txx} + 3uu_x = 2u_x u_{xx} + uu_{xxx}. \qquad (4.29)$$

As we know the solitary waves (peakons) are $u(t, x) = ce^{-|x - ct|}$, $c \in \mathbf{R}^1$, $c \neq 0$.

We shall put $\varphi(\lambda) = e^{-|\lambda|}$ and shall consider the Cauchy problem for (4.29) with initial condition $u(0, x) = u_0(x)$. Certainly, we must define the nonlinear terms in appropriate Sobolev spaces, say for $u_0 \in H^3$ let $u(t, x) \in C([0, T] : H^3(\mathbf{R}^1_x)) \cap C^1([0, T] : H^2(\mathbf{R}^1_x))$. Then we introduce the pseudodifferential operator $Q = (1 - \partial_x^2)^{1/2}$ having the symbol $q(\xi) = (1 + \xi^2)^{1/2}$. The operator $Q^{-2} = (1 - \partial_x^2)^{-1}$ is acting on the function $f \in L^2$ according to the formula $Q^{-2}f = p * f$, where $p(x) = \frac{1}{2}e^{-|x|} \in L^1$. In fact $(1 + \xi^2)^{-1}\hat{f} = \hat{p}\hat{f}$ and the Fourier transform $\hat{p}(\xi) = (1 + \xi^2)^{-1}$. One can easily see that (4.29) implies $Q^2(u_t + uu_x) = -\partial_x(u^2 + \frac{1}{2}u_x^2)$, i.e.

$$u_t + uu_x = -\partial_x \left(p * \left(u^2 + \frac{1}{2}u_x^2 \right) \right). \qquad (4.30)$$

Therefore,

$$u_t + \partial_x \left(\frac{1}{2} u^2 + p * \left[u^2 + \frac{1}{2} u_x^2 \right] \right) = 0 \qquad (4.31)$$

We remind of the reader that if $q, r \in H^s(\mathbf{R}^1)$, $s > \frac{1}{2}$, then $qr \in H^s(\mathbf{R}^1)$ and if $f_1 \in L_1$, $g_1 \in L_2$ then $f_1 * g_1 \in L_2$. The peakons should satisfy in distributional sense (4.31) as $u_t \in C([0,T) : H^2)$, $u^2 \in C([0,T) : H^3)$, $u_x^2 \in C([0,T) : H^2)$. In general, the nonlocal equation (4.31) holds in $C([0,T) : H^1)$. One could expect that a wave starting close to a solitary wave remains close to some translate of it all the later times. Roughly speaking, the profile of the wave remains almost the same for all times. In [34] a nice and simple proof of the following orbital stability result was given.

Theorem 4.5. *Let* $u \in C([0,T) : H^1(\mathbf{R}_x^1))$, $0 < T \leq \infty$ *be a generalized (weak) solution of (4.29), i.e. verifies (4.31) and such that* $\|u(0,.) - c\varphi\|_{H^1} < (\frac{\varepsilon}{3c})^2$ *for* $0 < \varepsilon < c$. *Then* $\|u(t,.) - c\varphi(. - \xi(t))\|_{H^1} < \varepsilon$ *for* $0 \leq t < T$, *where* $\xi(t) \in \mathbf{R}^1$ *is any point where the continuous function* $u(t,.)$ *attains its maximum.*

As we know, the Sobolev imbedding theorem asserts that for t fixed $u(t,x)$ is continuous with respect to x and tends to 0 at infinity. Thus, such a maximum exists and is attained.

We propose below a short sketch of the proof of this theorem as it can be useful in another situations. The proof will be divided into several steps.

Step 1. Differentiating (4.31) with respect to x, using the identities $\partial_x^2 = 1 - Q^2$, $Q^2(p * f) = f$, multiplying (4.31) by u and respectively $u_{tx} + u u_{xx} + u_x^2 - (u^2 + \frac{1}{2} u_x^2) + p * (u^2 + \frac{1}{2} u_x^2) = 0$ by u_x we get

$$\frac{1}{2}(u^2 + u_x^2)_t + \left[\frac{u u_x^2}{2} + u p * \left(u^2 + \frac{u_x^2}{2} \right) \right]_x = 0 \qquad (4.32)$$

for the smooth functions u.

Hence, the total energy $E(u)(t) = \int_{\mathbf{R}_x^1} (u^2 + u_x^2) dx = E(u_0)$ is constant in time. Similarly, $F(u)(t) = \int_{\mathbf{R}^1} (u^3 + u u_x^2) dx = F(u_0)$, $F(u) \leq M E(u)$, where $M = \max u$.

Step 2. To fix the ideas let $c = 1$. Then one can easily see that for each function $u \in H^1(\mathbf{R}_x^1)$ and $\xi \in \mathbf{R}^1$: $E(u) - E(\varphi) = \|u - \varphi(. - \xi)\|_{H^1}^2 + 4u(\xi) - 4$. In fact, $E(u) = \|u\|_{H^1}^2$ and the scalar product in H^1 is $(u,v)_{H^1} = \int_{\mathbf{R}^1} u(x)v(x) dx + \int_{\mathbf{R}^1} u_x v_x dx$, while $\|u - \varphi(. - \xi)\|_{H^1}^2 = \|u\|_{H^1}^2 + \|\varphi\|_{H^1}^2 - 2(u, \varphi(. - \xi))_{H^1}$, $\|\varphi\|_{H^1}^2 = 2$; $\varphi = e^{-|x|}$.

Step 3. Let $u \in H^1(\mathbf{R}^1)$, $M = max\, u$. Then

$$F(u) \leq ME(u) - \frac{2}{3}M^3.$$

The proof is simple. Denote $M = u(\xi)$ and introduce the function $g(x) = $
$$\begin{cases} u(x) - u_x(x), & x < \xi \\ u(x) + u_x(x), & x > \xi. \end{cases}$$
Then $\int_{\mathbf{R}} g^2(x)dx = E(u) - 2M^2$, $\int_{\mathbf{R}} u(x)g^2(x)dx = F(u) - \frac{4}{3}M^3 \Rightarrow$
$F(u) - \frac{4}{3}M^3 \leq M\int_{\mathbf{R}} g^2(x)dx.$

Step 4. Let $u \in H^1$ and $||u - \varphi||_{H^1} < \delta$. Then $|E(u) - E(\varphi)| \leq \delta(2\sqrt{2} + \delta)$, $|F(u) - F(\varphi)| \leq \delta(3\sqrt{2} + 3\delta + \frac{\delta^2}{\sqrt{2}})$. Certainly, $E(\varphi) = 2$, $F(\varphi) = \frac{4}{3}$. The proof of the first inequality is trivial, as for the second one, we use the identity

$$|F(u) - F(\varphi)| = \left| \int_{\mathbf{R}} (u - \varphi)(u^2 + u_x^2)dx + \int_{\mathbf{R}} \varphi(u^2 + u_x^2 - \varphi^2 - \varphi_x^2)dx \right|$$

$$= \left| \int_{\mathbf{R}} (u - \varphi)(u^2 + u_x^2)dx + \int_{\mathbf{R}} \varphi[(u - \varphi)^2 + (u_x - \varphi_x)^2]dx \right.$$

$$\left. + \int_{\mathbf{R}} 2\varphi[(u - \varphi)\varphi + (u_x - \varphi_x)\varphi_x]dx \right|$$

and the Sobolev embedding theorem for $v \in H^1(\mathbf{R}) : |v(x)| \leq \frac{1}{\sqrt{2}}||v||_{H^1}$.

Step 5. Let $u \in H^1(\mathbf{R})$, $M = max\, u$ and $|E(u) - E(\varphi)| \leq \delta(2\sqrt{2} + \delta)$, $|F(u) - F(\varphi)| \leq \delta(3\sqrt{2} + 3\delta + \frac{1}{\sqrt{2}}\delta^2)$ for $0 < \delta < \frac{1}{20}$. Then $|M - 1| \leq 2\sqrt{\delta}$. The proof is based on the properties of the cubic polynomials (continuous dependence of the roots of the coefficients etc.) More precisely,

$$E(u) \sim E(\varphi) = 2, F(u) \sim F(\varphi) = \frac{4}{3} \Rightarrow M \geq \frac{F(u)}{E(u)} > \frac{1}{2}.$$

According to Step 3: $M^3 - \frac{3}{2}ME(u) + \frac{3}{2}F(u) \leq 0$, i.e. we obtain a cubic polynomial with respect to $M \to y$, whose coefficients depend continuously on $E(u)$ and $F(u)$: $P(y) = y^3 - \frac{3}{2}yE(u) + \frac{3}{2}F(u)$. In the special case $E(u) = E(\varphi) = 2$, $F(u) = F(\varphi) = \frac{4}{3}$, $P(y)$ takes the form $P_0(y) = y^3 - 3y + 2 = (y - 1)^2(y + 2)$. Therefore P has a simple root near $y = -2$ and no more than two roots (eventually double), near 1. As $M > \frac{1}{2}$ we conclude that $M \approx 1$. Technical details are omitted.

Proof of Theorem 4.5. Let $u_0 = u(0, x)$ is such that $||u_0 - \varphi||_{H^1} < \frac{\varepsilon^4}{84}$, $0 < \varepsilon < 1$. Put $\delta = \frac{\varepsilon^4}{81} \Rightarrow 0 < \delta < \frac{1}{20}$. According to the result formulated in Step 4: $|E(u_0) - E(\varphi)| \leq \delta(2\sqrt{2} + \delta)$, $|F(u_0) - \frac{4}{3}| \leq \delta(3\sqrt{2} + 3\delta + \frac{1}{\sqrt{2}}\delta^2)$.

Having in mind that $E(u) = E(u_0)$, $F(u) = F(u_0)$ we can apply Step 5 and conclude that $|u(t, \xi(t)) - 1| \leq 2\frac{\varepsilon^2}{9}$, $\forall t \in [0, T)$. Consequently, $||u(t, .) - \varphi(. - \xi(t))||^2_{H^1} = E(u_0) - E(\varphi) + 4(1 - u(t, \xi(t))$ as $u(t, \xi(t)) = M$, i.e. $||u(t, .) - \varphi(. - \xi(t))||^2_{H^1} \leq \varepsilon^2$, $\forall t \in (0, T)$.

It is interesting to discuss the problem about globality of u, i.e. $T = \infty$. Suppose that $u_0 \in H^3(\mathbf{R})$ and $y_0 = u_0 - u_{0xx}$ does not change sign. Then Constantin and Escher [29], [30], [31] proved that there exists a unique global solution $u \in C([0, \infty) : H^3(\mathbf{R})) \cap C^1([0, \infty) : H^2(\mathbf{R}))$. Wave breaking occurs if (say) either u_0 is odd and $u'_0(0) < 0$ or $u'_0(x_0) < -\frac{1}{\sqrt{2}}||u_0||_{H^1}$.

Depending on the velocity c ($c > 0$, $c < 0$) the Camassa-Holm equation has peaked solitary waves that can travel on both directions of the real line. The positive ones travel to the right (peakons), while the negative ones travel to the left (antipeakons). In [46] the stability of ordered trains of antipeakons and peakons is studied. Interesting papers on the subject appeared in [120], [32], [20].

Chapter 5

Interaction of two peakons satisfying the Camassa-Holm equation

5.1 Introduction. Construction of two-peakon solutions

1. Consider the Camassa-Holm equation written in the form:

$$u_t - u_{xxt} + 3uu_x = \frac{d}{dx}\left(\frac{1}{2}u_x^2 + uu_{xx}\right). \tag{5.1}$$

We are looking for a distribution solution of (5.1) having the form

$$u(t,x) = p_1(t)e^{-|x-q_1(t)|} + p_2(t)e^{-|x-q_2(t)|}, \tag{5.2}$$

p_1, p_2, q_1, q_2 being classical C^1 functions. Certainly, the two waves participating in (5.2) are peakons (two-peakon solutions).

The definition of distribution (weak) solution of (5.1) and its properties are discussed into details in [29] (see the Footnote of Chapter IV). For the further development of the same subject see [11], [12], [19], [20]. We omit that facts in Chapter V as we are concentrating on the interaction of two peakons proposing here all the necessary technical specifictaions. Moreover, our considerations are not complicated as they are based on some classical results from Schwartz distribution theory and on the Hamiltonian systems. Several geometrical illustrations simplify the things. The Ansatz (5.2) is motivated by Theorem 1.1 (1). One can easily see that the peakon solution constructed there satisfies (5.1) in a weak sense (see [4], [31]). We must find the phase and amplitude functions q_1, q_2, p_1, p_2 in such a way that (5.2) verifies (5.1) in a distribution sense.

2. The above mentioned problem is not trivial as in general multiplication of distributions is not possible and (5.1) is a quasilinear equation. To simplify the things we suppose that q_1, q_2, p_1, $p_2 \in C^1$ and $q_1(t) \neq q_2(t)$, $\forall t$. To begin with we split $\mathbf{R}^2$ into 3 parts (see Fig. 5.1), where

I: $x > q_1(t)$, II: $q_1(t) > x > q_2(t)$, III: $x < q_2(t)$ and $\Gamma_1 : x = q_1(t)$, $\Gamma_2 : x = q_2(t)$.

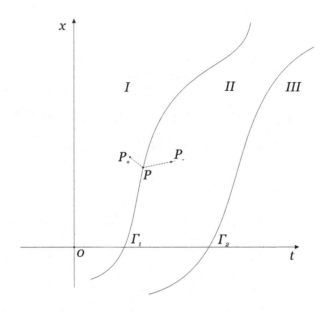

Fig. 5.1

Evidently, in I: $|x - q_1(t)| = x - q_1(t)$, $|x - q_2(t)| = x - q_2(t)$; in II: $|x - q_1(t)| = q_1(t) - x$, $|x - q_2(t)| = x - q_2(t)$; in III: $|x - q_1(t)| = q_1(t) - x$, $|x - q_2(t)| = q_2(t) - x$. In other words, in I : $u = p_1 e^{q_1 - x} + p_2 e^{q_2 - x}$ $\Rightarrow u_t = (p_1' + p_1 q_1')e^{q_1 - x} + (p_2' + p_2 q_2')e^{q_2 - x}$, $u_{xxt} = u_t$, $u_x = -u$, $u_{xx} = -u_x = u$, $u_{xxx} = -u$. Thus, (5.2) satisfies (5.1) in I and evidently in III. As it concerns II $\Rightarrow u = p_1 e^{x - q_1} + p_2 e^{q_2 - x} \Rightarrow u_t = (p_1' - p_1 q_1')e^{x - q_1} + (p_2' + p_2 q_2')e^{q_2 - x} \Rightarrow u_{txx} = u_t$; $u_x = p_1 e^{x - q_1} - p_2 e^{q_2 - x} \Rightarrow u_{xx} = u$, $u_{xxx} = u_x$. Thus, (5.2) satisfies in classical sense (5.1) in II.

Conclusion: The function u given by (5.2) satisfies (5.1) in classical sense outside $\Gamma_1 \cup \Gamma_2$. (i.e. in $I \cup II \cup III$).

We are interested now what will happen near Γ_1 and Γ_2. To do this we shall use the so called jump formula from the theory of Schwartz distributions. We shall remind of the reader the corresponding formula ([52], [117]). Thus, let $f \in C^1(\bar{I}) \cap C^1(\overline{II \cup III})$, where $\bar{A}$ is the closure of the set A. Then

$$\frac{\partial f}{\partial t} = \left\{ \frac{\partial f}{\partial t} \right\} - q_1'(t)[f]_{\Gamma_1}\delta_{\Gamma_1} \tag{5.3}$$

$$\frac{\partial f}{\partial x} = \left\{ \frac{\partial f}{\partial x} \right\} + [f]_{\Gamma_1}\delta_{\Gamma_1}, \tag{5.4}$$

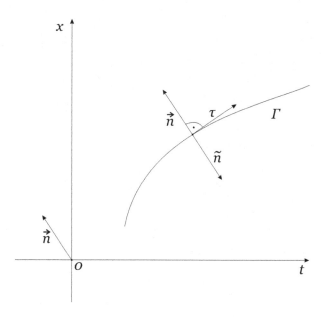

Fig. 5.2

where $\frac{\partial f}{\partial t} \in D'(\mathbf{R}^2)$ is the Schwartz generalized derivative of $f \in D'(\mathbf{R}^2)$, $\{\frac{\partial f}{\partial t}\}$ means the regular distribution (function) generated by $\frac{\partial f}{\partial t} \in C^0(I) \cap C^0(I \cup II)$, i.e. the function $\{\frac{\partial f}{\partial t}\}$ is defined almost everywhere (outside Γ_1), $[f]_{\Gamma_1}$ is the jump of f on Γ_1, i.e. $[f]_{\Gamma_1}(P) = lim_{P^+ \to P} f(P^+) - lim_{P^- \to P} f(P^-)$, where $P \in \Gamma_1$, $P^+ \in I$, $P^- \in II \cup III$ and $\Gamma_1 : \left|\begin{smallmatrix} t = t \\ x = q_1(t) \end{smallmatrix}\right.$, δ_{Γ_1} is the standard δ function on Γ_1 see [57]), i.e. $\forall \varphi \in C_0^\infty(\mathbf{R}^2)$:

$$(\delta_{\Gamma_1}, \varphi) = \int_{\Gamma_1} \varphi(t, q_1(t)) dt$$

(see Fig. 5.2).

Certainly, $|x| = x \, sgn \, x$ and jump formulas similar to (5.3), (5.4) are valid for $g \in C^1(\overline{I \cup II}) \cap C^1(\overline{III})$ too. Thus, applying (5.3), (5.4) to (5.2) we get $u_t, u_x \in D'$,

$$u_t = \{u_t\}, u_x = \{u_x\} \tag{5.5}$$

as u is continuous in $\mathbf{R}^2$.

Evidently,

(*) $u_x = \{-p_1\, sgn(x - q_1)e^{-|x-q_1|} - p_2\, sgn(x - q_2)e^{-|x-q_2|}\}$

outside Γ_1, Γ_2, as $sgn(x - q_1) = sgn(x - q_2) = 1$ in I, $sgn(x - q_1) = -1$, $sgn(x - q_2) = 1$ in II, $sgn(x - q_1) = -1$, $sgn(x - q_2) = 1$ in III. Then we have that almost everywhere, i.e. in $\mathbf{R}^2 \setminus (\Gamma_1 \cup \Gamma_2)$:

$$u_x^2 = \{u_x^2\} = \{p_1^2 e^{-2|x-q_1|} + 2p_1 p_2\, sgn(x - q_1)$$

$$\times\, sgn(x - q_2)e^{-|x-q_1|-|x-q_2|} + p_2^2 e^{-2|x-q_2|}$$

and $p_1^2 e^{-2|x-q_1|}$, $p_2^2 e^{-2|x-q_2|}$ are continuous in $\mathbf{R}^2$, while $2p_1 p_2\, sgn(x - q_1)sgn(x - q_2)e^{-|x-q_1|-|x-q_2|}$ has jumps on Γ_1 and Γ_2.

Thus, in $D'(\mathbf{R}^2)$ we have:

$$\frac{1}{2}\frac{\partial}{\partial x}u_x^2 = \left\{\frac{1}{2}\frac{\partial}{\partial x}u_x^2\right\} + [u_x^2]_{\Gamma_1}\delta_{\Gamma_1} + [u_x^2]_{\Gamma_2}\delta_{\Gamma_2}$$

$$= \left\{\frac{1}{2}\frac{\partial}{\partial x}u_x^2\right\} + 2p_1 p_2\, sgn(q_1 - q_2)e^{-|q_1-q_2|}$$

$$\times\, \delta_{\Gamma_1} + 2p_1 p_2\, sgn(q_2 - q_1)e^{-|q_1-q_2|}\delta_{\Gamma_2}.$$

(We use the fact that in $II : q_1 > q_2$.)

Evidently,

$$3uu_x = \{3uu_x\} \tag{5.6}$$

as $u \in C^0(\mathbf{R}^2)$.

Having in mind that

$$u_t = \{(p_1' + p_1 q_1'\, sgn(x - q_1))e^{-|x-q_1|}$$

$$+ (p_2' + p_2 q_2'\, sgn(x - q_2))e^{-|x-q_2|}\} - q_1'[u]_{\Gamma_1}\delta_{\Gamma_1} - q_2'[u]_{\Gamma_2}\delta_{\Gamma_2}$$

and $[u]_{\Gamma_1} = [u]_{\Gamma_2} = 0$ we see that

$$u_{tx} = \{u_{tx}\} + [u_t]_{\Gamma_1}\delta_{\Gamma_1} + [u_t]_{\Gamma_2}\delta_{\Gamma_2} = \{u_{tx}\} + 2p_1 q_1'\delta_{\Gamma_1} + 2p_2 q_2'\delta_{\Gamma_2}. \tag{5.7}$$

Differentiating (5.7) in D' and applying again (5.4) we have:

$$u_{txx} = \frac{\partial}{\partial x}\{u_{tx}\} + 2p_1 q_1'\frac{\partial}{\partial x}\delta_{\Gamma_1} + 2p_2 q_2'\frac{\partial}{\partial x}\delta_{\Gamma_2} \tag{5.8}$$

$$= \{u_{txx}\} + [u_{tx}]_{\Gamma_1}\delta_{\Gamma_1} + [u_{tx}]_{\Gamma_2}\delta_{\Gamma_2} + 2p_1 q_1'\frac{\partial}{\partial x}\delta_{\Gamma_1} + 2p_2 q_2'\frac{\partial}{\partial x}\delta_{\Gamma_2}.$$

On the other hand, (*) implies that outside $\Gamma_1 \cup \Gamma_2$:

$$\{u_{tx}\} = \{(-p_1'\, sgn(x-q_1)+p_1 q_1')e^{-|x-q_1|}+(-p_2'\, sgn(x-q_2)+p_2 q_2')e^{-|x-q_2|}\}$$

and therefore the corresponding jumps on Γ_1, Γ_2 are:

$$[u_{tx}]_{\Gamma_1} = -2p_1', [u_{tx}]_{\Gamma_2} = -2p_2'.$$

Thus (5.8) is rewritten as:

$$u_{txx} = \{u_{txx}\} - 2p_1'\delta_{\Gamma_1} - 2p_2'\delta_{\Gamma_2} + 2p_1q_1'\frac{\partial}{\partial x}\delta_{\Gamma_2} + 2p_2q_2'\frac{\partial}{\partial x}\delta_{\Gamma_2}. \quad (5.9)$$

To complete the things, we observe that in D':

$$\partial_x(u_x) = u_{xx} = \{u_{xx}\} + [u_x]_{\Gamma_1}\delta_{\Gamma_1} + [u_x]_{\Gamma_2}\delta_{\Gamma_2} \quad (5.10)$$

$$= \{u_{xx}\} - 2p_1\delta_{\Gamma_1} - 2p_2\delta_{\Gamma_2} \Rightarrow uu_{xx} = \{uu_{xx}\} - 2p_1u|_{\Gamma_1}\delta_{\Gamma_1} - 2p_2u|_{\Gamma_2}\delta_{\Gamma_2}$$

$$= \{uu_{xx}\} - 2p_1(p_1 + p_2e^{-|q_1-q_2|})\delta_{\Gamma_1} - 2p_2(p_2 + p_1e^{-|q_1-q_2|})\delta_{\Gamma_2}$$

as $u \in C^0(\mathbf{R}^2)$.

Therefore, differentiating in D' we get:

$$\frac{\partial}{\partial x}(uu_{xx}) = \{(uu_{xx})_x'\} + [uu_{xx}]_{\Gamma_1}\delta_{\Gamma_1} + [uu_{xx}]_{\Gamma_2}\delta_{\Gamma_2} \quad (5.11)$$

$$-2p_1(p_1 + p_2e^{-|q_1-q_2|})\frac{\partial\delta_{\Gamma_1}}{\partial x} - 2p_2(p_2 + p_1e^{-|q_1-q_2|})\frac{\partial\delta_{\Gamma_2}}{\partial x}.$$

Evidently, $u \in C^0(\mathbf{R}^2) \Rightarrow [uu_{xx}]_{\Gamma_{1,2}} = u|_{\Gamma_{1,2}}[u_{xx}]_{\Gamma_{1,2}}$. From (*) we get

$$\{u_{xx}\} = \{+p_1e^{-|x-q_1|} + p_2e^{-|x-q_2|}\} = \{u\} \Rightarrow [uu_{xx}]_{\Gamma_{1,2}} = 0.$$

Combining (5.5)–(5.11) and substituting in (5.1) we obtain:

$$u_t - u_{xxt} + 3uu_x - \frac{\partial}{\partial x}\left(\frac{u_x^2}{2} + uu_{xx}\right)$$

$$= \left\{u_t - u_{xxt} + 3uu_x - \frac{\partial}{\partial x}\left(\frac{u_x^2}{2} + uu_{xx}\right)\right\}$$

$$+ 2p_1'\delta_{\Gamma_1} + 2p_2'\delta_{\Gamma_2} - 2p_1q_1'\frac{\partial}{\partial x}\delta_{\Gamma_1} - 2p_2q_2'\frac{\partial}{\partial x}\delta_{\Gamma_2}$$

$$- 2p_1p_2\,sgn(q_1 - q_2)e^{-|q_1-q_2|}\delta_{\Gamma_1} - 2p_1p_2\,sgn(q_2 - q_1)e^{-|q_1-q_2|}\delta_{\Gamma_2}$$

$$+ 2p_1(p_1 + p_2e^{-|q_1-q_2|})\frac{\partial\delta_{\Gamma_1}}{\partial x} + 2p_2(p_2 + p_1e^{-|q_1-q_2|})\frac{\partial\delta_{\Gamma_2}}{\partial x}.$$

Therefore, (5.1) is satisfied in D' (in weak sense) by Ansatz (5.2) iff:

$$\begin{vmatrix} p_1' - p_1p_2\,sgn(q_1 - q_2)e^{-|q_1-q_2|} = 0 \\ p_2' - p_1p_2\,sgn(q_2 - q_1)e^{-|q_1-q_2|} = 0 \\ (p_1 + p_2e^{-|q_1-q_2|}) - q_1' = 0 \\ (p_2 + p_1e^{-|q_1-q_2|}) - q_2' = 0 \end{vmatrix} \quad (5.12)$$

(5.12) is a system of ODE but having discontinuous right hand sides ($sgn(q_1 - q_2)$ is not C^0 function for $q_1(t) = q_2(t)$). The system (5.12) of Hamiltonian type appears at first in paper [29]. In its investigation given below we shall follow closely [29]. We shall have to overthrow many difficulties when solving (5.12). For example, we can have non unicity results etc. There are no difficulties if we guarantee that $q_1(t) \neq q_2(t)$ everywhere. Then either $q_1(t) > q_2(t), \forall t$ or $q_1(t) < q_2(t), \forall t$.

To continue our work we observe that the traveling wave (peakon, soliton) $v_c(t, x) = ce^{-|x - ct|}$, $c = const$ is for any $c > 0$ a distribution solution of (5.1) (see Theorem 4.5). It is interesting to note that the speed of the wave is equal to its amplitude and that v_c has a corner point at its peak $x = ct$.

5.2   Interaction of two peakons via the study of the system (5.12)

1. We shall suppose now that the soliton waves $\tilde{I} : p_1(t)e^{-|x - q_1(t)|}$ and $\tilde{II} : p_2(t)e^{-|x - q_2(t)|}$ are separated at $t = -\infty$ and that they have asymptotic speeds and amplitudes at $t = -\infty$, $c_1 > 0$ and $c_2 > 0$ respectively, $c_1 > c_2 > 0$. Thus, the collision between them occurs if the faster wave (with speed $c_1 > 0$) is located to the left of the slower (with speed $c_2 > 0$). So we assume that $p_1(t) \to c_1$, $p_2(t) \to c_2$, $q_1(t) - c_1 t \to 0$, $q_2 - c_2 t \to 0$ for $t \to -\infty$, i.e. $u(t, x) \approx c_1 e^{-|x - c_1 t|} + c_2 e^{-|x - c_2 t|}$ as $t \to -\infty$.

The standard change

$$P = p_1 + p_2, Q = q_1 + q_2 \tag{5.13}$$

$$p = p_1 - p_2, q = q_1 - q_2$$

in the ODE system (5.12) leads to the following equivalent Hamiltonian system in the new variables $(q, Q; p, P)$:

$$\left| \begin{array}{l} q' = \dfrac{\partial H}{\partial p} = p(1 - e^{-|q|}), p' = -\dfrac{\partial H}{\partial q} = \dfrac{1}{2}(P^2 - p^2)e^{-|q|}\, sgn\, q \\[2mm] Q' = \dfrac{\partial H}{\partial P} = P(1 + e^{-|q|}), P' = -\dfrac{\partial H}{\partial Q} = 0, \end{array} \right. \tag{5.14}$$

where the Hamiltonian $H = \frac{1}{2}P^2(1 + e^{-|q|}) + \frac{1}{2}p^2(1 - e^{-|q|})$; $\frac{\partial}{\partial q}e^{-|q|}) = -e^{-|q|}\, sgn\, q$.

One can easily see that $P \equiv c_1 + c_2$, $q(t) - (c_1 - c_2)t \to 0$, $t \to -\infty$; $p(t) \to c_1 - c_2$, $t \to -\infty$; $Q(t) - (c_1 + c_2)t \to 0$ for $t \to -\infty$, i.e. $q(t) \to -\infty$

and $Q(t) \to -\infty$ for $t \to -\infty$. As the Hamiltonian is a first integral of (5.14) we have that along the phase curve of (5.14) under consideration $H(q, Q; p, P) = const = \frac{1}{2}[(c_1 + c_2)^2 + (c_1 - c_2)^2] = c_1^2 + c_2^2 > 0$. Thus,

$$e^{-|q|} = \frac{2H - c^2 - p^2}{c^2 - p^2},$$

where $c = c_1 + c_2$ and we shall denote $L = c_1 - c_2 > 0$.

Suppose now that for some t^* the peaks of $\tilde{I}$ and $\tilde{II}$ overlap, i.e. $x = q_1(t^*)$, $x = q_2(t^*) \Rightarrow q(t^*) = 0 \Rightarrow c^2 = 2H - c^2 \Rightarrow H = c^2 = (c_1 + c_2)^2$ - contradiction with $H = c_1^2 + c_2^2$. Geometrically, such a situation is impossible.

Hence, $q(t) \neq 0$ everywhere and therefore either $q(t) > 0, \forall t$ or $q(t) < 0$ for each t. The fact that $lim_{t \to -\infty} q(t) = -\infty \Rightarrow q(t) < 0$ everywhere. Consequently, $sgn\, q = -1$, $e^{-|q|} = e^q \in (0, 1)$ and the solutions of (5.14) are unique and C^∞ smooth. This way we avoid the main difficulty in studying (5.14) due to the fact that the right hand side in (5.14) is discontinuous in general ($sgn\, q$ is not continuous near $q(t_0) = 0$). From (5.14) we get $p' = \frac{1}{2}(p^2 - L^2) \Rightarrow \frac{2dp}{p^2 - L^2} = dt \Rightarrow (\frac{1}{p-L} - \frac{1}{p+L})dp = Ldt \Rightarrow ln|\frac{p-L}{p+L}| = Lt + k$, $k = ln|\frac{p(0)-L}{p(0)+L}|$.

Evidently, (5.14) implies that

$$p' = \frac{1}{2}(c^2 - p^2)\left(-\frac{d}{dq}e^q\right) \Rightarrow dp = \frac{1}{2}(c^2 - p^2)d(1 - e^q)\frac{dt}{dq}$$

and $dq = p(1 - e^q)dt$. Therefore,

$$2pdp = (c^2 - p^2)d(1 - e^q)\frac{1}{(1 - e^q)}$$

$$\Rightarrow \frac{2pdp}{c^2 - p^2} = d\,ln(1 - e^q)(0 < e^q < 1).$$

Consequently,

$$-\frac{d(p^2 - c^2)}{c^2 - p^2} = -d\,ln(1 - e^q) \Rightarrow +ln|p^2 - c^2| = -ln(1 - e^q) + ln\,\tilde{c}, \tilde{c} > 0,$$

i.e. $|p^2 - c^2| = \frac{\tilde{c}}{1-e^q} > 0$.

In other words either $p^2(t) > c^2$ everywhere or $p^2(t) < c^2$. Having in mind that $p^2(-\infty) = L^2 > 0$, $0 < L < c$ we get that $c^2 > p^2(t), \forall t$. Then the ODE $p' = -\frac{1}{2}(c^2 - p^2)e^q$ implies that $p'(t) < 0$ everywhere, i.e. $p(t)$ is strictly monotonically decreasing function on $\mathbf{R}^1$, $lim_{t \to -\infty} p(t) = L$ (see Fig. 5.3).

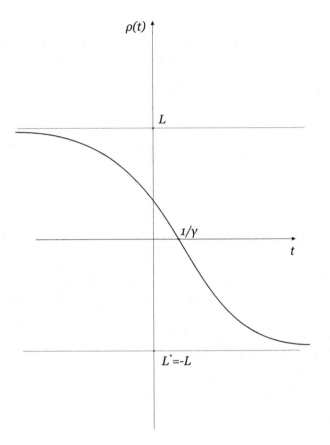

Fig. 5.3

Then $|\frac{p-L}{p+L}| = e^k e^{Lt} = \gamma e^{Lt}$, $\gamma = e^k > 0$. Thus $-L = L^* < p(t) < L \Rightarrow$ $\frac{L-p}{p+L} = \gamma e^{Lt} \Rightarrow$

$$p(t) = L\frac{1 - \gamma e^{Lt}}{1 + \gamma e^{Lt}}, \forall t \in \mathbf{R}^1. \tag{5.15}$$

The equation $p' = \frac{1}{2}(p^2 - c^2)e^q$ and (5.15) give us

$$q(t) = ln\frac{L^2\gamma e^{Lt}}{(c_1\gamma e^{Lt} + c_2)(c_2\gamma e^{Lt} + c_1)}, t \in \mathbf{R}^1. \tag{5.16}$$

From (5.16) we can find the asymptotic behavior of $q(t)$ at $\pm\infty$:

$$lim_{t\to\infty}(q(t) + Lt) = lim_{t\to\infty} ln\frac{L^2\gamma e^{2Lt}}{(c_1\gamma e^{Lt} + c_2)(c_2\gamma e^{Lt} + c_1)}$$

$$= ln\frac{L^2}{c_1c_2\gamma} = ln\frac{L^2}{c_1c_2} - ln\gamma \quad as \quad ln\,e^{Lt} = Lt;$$

$$lim_{t\to-\infty}(q(t) - Lt) = lim_{t\to-\infty} ln\frac{L^2\gamma}{(c_1\gamma e^{Lt} + c_2)(c_2\gamma e^{Lt} + c_1)}$$

$$= ln\frac{L^2\gamma}{c_1c_2} = ln\frac{L^2}{c_1c_2} + ln\,\gamma.$$

On the other hand,

$$lim_{t\to-\infty}(q(t) - Lt) = 0 \Rightarrow \gamma = \frac{c_1c_2}{L^2}$$

$$\Rightarrow lim_{t\to\infty}(q(t) + Lt) = 2\,ln\frac{L^2}{c_1c_2}. \tag{5.17}$$

Our last technical step is to find $Q(t)$. From (5.14): $Q' = c(1 + e^q) = c(1 + \frac{L^2\gamma e^{Lt}}{(c_1\gamma e^{Lt}+c_2)(c_2\gamma e^{Lt}+c_1)})$, i.e. $Q(t) = Q(0) + ct + cL^2\gamma \int_0^t \frac{e^{Ls}ds}{(c_1\gamma e^{Ls}+c_2)(c_2\gamma e^{Ls}+c_1)}$.

There are no difficulties to compute

$$cL^2\gamma \int \frac{e^{Lt}dt}{(c_1\gamma e^{Lt} + c_2)(c_2\gamma e^{Lt} + c_1)}$$

$$= ln\frac{(c_1\gamma e^{Lt} + c_2)(c_2\gamma + c_1)}{(c_2\gamma e^{Lt} + c_1)(c_1\gamma + c_2)} = \Phi(t), \Phi(0) = 0.$$

Thus,

$$Q(t) = Q(0) + ct + ln\frac{(c_1\gamma e^{Lt} + c_2)(c_2\gamma + c_1)}{(c_2\gamma e^{Lt} + c_1)(c_1\gamma + c_2)} \tag{5.18}$$

and consequently,

$$lim_{t\to\infty}(Q(t) - ct) = Q(0) + lim_{t\to\infty} ln\frac{(c_1\gamma e^{Lt} + c_2)(c_2\gamma + c_1)}{(c_2\gamma e^{Lt} + c_1)(c_1\gamma + c_2)}$$

$$= Q(0) + ln\frac{c_1(c_2\gamma + c_1)}{c_2(c_1\gamma + c_2)}.$$

On the other hand,

$$0 = lim_{t\to-\infty}(Q(t) - ct) = Q(0) + ln\frac{c_2(c_2\gamma + c_1)}{c_1(c_1\gamma + c_2)}$$

according to (5.18). So, $Q(0) = ln\frac{c_1(c_1\gamma+c_2)}{c_2(c_2\gamma+c_1)}$.

Therefore,

$$lim_{t\to\infty}(Q(t) - ct) = ln\left(\frac{c_1.c_1}{c_2.c_2}\right) = 2\,ln\frac{c_1}{c_2}. \qquad (5.19)$$

Combining $p_1 + p_2 = c = c_1 + c_2$ and (5.15): $lim_{t\to\infty}p(t) = lim_{t\to\infty}(p_1 - p_2) = -L = -c_1 + c_2 < 0$, as well as (5.17), (5.19) we get: $lim_{t\to\infty}p_1(t) = c_2$, $lim_{t\to\infty}p_2(t) = c_1$, $lim_{t\to\infty}(q_1(t) - c_2t) = 2\,ln\frac{L}{c_2}$, $lim_{t\to\infty}(q_2(t) - c_1t) = 2\,ln\frac{c_1}{L}$.

Conclusion:

$$u(t,x) \approx c_2 e^{|x-c_2t-2\,ln\frac{L}{c_2}|} + c_1 e^{|x-c_1t-2\,ln\frac{c_1}{L}|} \qquad (5.20)$$

for $t \to \infty$, i.e.

$$u(t,x) \approx c_1 e^{-|x-c_1t-2\,ln\frac{c_1}{c_1-c_2}|} + c_2 e^{-|x-c_2t-2\,ln\frac{c_1-c_2}{c_2}|}, t \to \infty. \qquad (5.21)$$

2. We propose now the geometrical interpretation of (5.21). The taller wave $\tilde{I}$: $p_1(t)e^{-|x-q_1(t)|}$ starting to the left of the shorter one $\tilde{II}$: $p_2(t)e^{-|x-q_2(t)|}$ catches the shorter one and then they collide. For $t << -1 : \tilde{I} \approx c_1 e^{-|x-c_1t|}$, $\tilde{II} \approx c_2 e^{-|x-c_2t|}$ (see Fig. 5.4).

No overlapping of the peaks occurs.

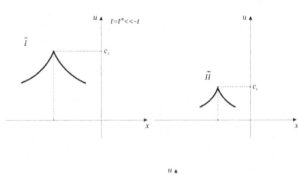

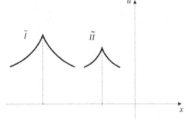

Fig. 5.4

We summarize below the main observations from this Chapter.

(i) if $c_1 > 2c_2 > 0 \Rightarrow ln\frac{c_1}{c_1-c_2} > 0$, $ln\frac{c_1-c_2}{c_2} > 0$ and both waves in (5.21) have a forward shift for $t >> 1$.

(ii) $c_1 = 2c_2 \Rightarrow ln\frac{c_1}{c_1-c_2} = ln2 > 0$, $ln\frac{c_1-c_2}{c_2} = ln1 = 0$ and no shift occurs for the shorter wave, while the taller is shifted forward.

(iii) $1 < c_1 < 2c_2 \Rightarrow ln\frac{c_1}{c_1-c_2} > 0$, $ln\frac{c_1-c_2}{c_2} < 0$, i.e. the taller wave is shifted forward, while the shorter wave is shifted backward. After the collision the taller wave reappears to the right of the shorter one. Due to the interaction (nonlinear effect) a phase shift appears. Geometrically see Figs. 5.5–5.8.

After the collision the two waves reemerge with their initial shapes and speeds. In other words, u consists of two soliton type solutions retaining their identities after the interaction. The peakons have a similar behaviour to the smooth soliton waves. This fact motivates the name of the Ansatz (5.2): two soliton-peakon solution of the Camassa-Holm equation (5.1).

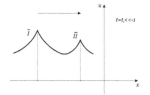

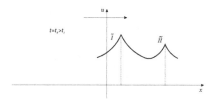

Fig. 5.5   Fig. 5.6

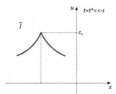

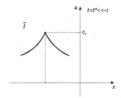

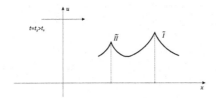

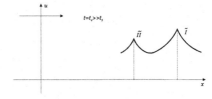

Fig. 5.7   Fig. 5.8

Chapter 6

Method of the characteristics applied to the 2nd and 3rd order Hunter-Saxton equation

6.1 Classical solutions of the Cauchy problem

1. At the beginning of this Chapter we shall discuss briefly the chemical interpretation of the so called Hunter-Saxton equation. As it is known the director field of a nematic liquid crystal satisfies some nonlinear wave equation. By definition nematic means that the corresponding crystal is characterized by having the long axes of the molecules in parallel lines but not layers. The orientation of the molecules is described by a field of unit vectors $(cos\ u(t,x), sin\ u(t,x)) = n(t,x) \in S^2$. As the nematic crystals are invariant under the mapping $n \to -n$ the vector field n is called a director field. In the simplest situation the dynamics of n is independent of any coupling of the fluid flow. It was shown in [58] that the weakly nonlinear unidirectional waves satisfy asymptotically (in precise mathematical sense) the second order Hunter-Saxton equation, where x is the space variable in a frame moving with the linearized wave velocity and t is a "slow time variable". The Cauchy problem for it can be formulated as follows:
Find a function $u \in C^2(t \geq 0)$ and such that

$$(u_t + uu_x)_x = \frac{1}{2}u_x^2, t \geq 0 \tag{6.1}$$

$$u(0,x) = F(x).$$

The function $F \in C^2(\mathbf{R}^2)$ and is bounded with its derivatives up to order 2, $\int_{-\infty}^{\infty}(F')^2(x)dx < \infty$, $F(x)_{x\to\infty} \to 0$. Suppose now that $u \in C^3(t \geq 0)$. By differentiating (6.1) with respect to x we obtain the third order Hunter-Saxton equation:

$$u_{txx} + 2u_x u_{xx} + uu_{xxx} = 0, t \geq 0, \tag{6.2}$$

$$u(0, x) = F(x).$$

Conversely, assume that $u \in C^3(t \geq 0)$ satisfies (6.2). Then (6.2) can be written in the form

$$\frac{\partial}{\partial x} \left(u_{tx} + \frac{1}{2}u_x^2 + uu_{xx} \right) = 0,$$

i.e. there exists a function $\varphi \in C^1(t \geq 0)$ and such that $u_{tx} + \frac{1}{2}u_x^2 + uu_{xx} = \varphi(t)$ or equivalently

$$(u_t + uu_x)_x - \frac{1}{2}u_x^2 = \varphi(t), t \geq 0 \tag{6.3}$$

$$u(0, x) = F(x).$$

Consider (6.3) with the right-hand side $\varphi \in C^1(t \geq 0)$. If φ is arbitrary, we can not expect the existence of a unique classical solution of (6.2) because (6.3) $\Rightarrow$ (6.2) after differentiation with respect to x. Because of this reason we shall concentrate only on the existence of a classical solution to the Cauchy problems (6.2), (6.3). Our study is based on [58], [36], [37].

2. As we know, the equation of the characteristics for (6.3) is given by

$$\frac{dx}{dt} = u(t, x) \tag{6.4}$$

$$x(0, \alpha) = \alpha \in \mathbf{R}^1,$$

i.e. (t, x) are the old coordinates, $(t, x(t, \alpha)) = (t, \alpha)$ are the new curvilinear coordinates, $\mathcal{L} : x = x(t, \alpha)$ are the characteristics, $\alpha \in \mathbf{R}^1$. Thus, $\frac{\partial x}{\partial t} = \tilde{u}(t, \alpha)$, where $\tilde{u}(t, \alpha) = u(t, x(t, \alpha)) = u|_{\mathcal{L}}$ and according to (6.4) $\frac{\partial \tilde{u}}{\partial t} = \frac{\partial u}{\partial t} + \frac{\partial u}{\partial x}\frac{dx}{dt} = u_t + uu_x = \tilde{w}$. Then in the old coordinates $u_t + uu_x = w \Rightarrow \frac{\partial w}{\partial x} = (u_t + uu_x)_x = \frac{\partial \tilde{w}}{\partial t}\frac{\partial t}{\partial x} + \frac{\partial \tilde{w}}{\partial \alpha}\frac{\partial \alpha}{\partial x} = \frac{\partial \tilde{w}}{\partial \alpha}\frac{\partial \alpha}{\partial x}$, because $\tilde{w}(t, \alpha) = w(t, x(t, \alpha))$.

Therefore, $\frac{\partial}{\partial x}(u_t + uu_x) = \frac{\partial^2 \tilde{u}}{\partial t \partial \alpha}\frac{\partial \alpha}{\partial x}$. Put $V(t, \alpha) = \frac{\partial x}{\partial \alpha} \Rightarrow \frac{\partial \alpha}{\partial x} = \frac{1}{V}$ $\Rightarrow \frac{\partial V}{\partial t} = \frac{\partial^2 x}{\partial t \partial \alpha} = \frac{\partial \tilde{u}}{\partial \alpha}$ from (6.4). Moreover, $u_x = \frac{\partial \tilde{u}}{\partial \alpha}\frac{\partial \alpha}{\partial x}$.

So we conclude that in the new coordinates (t, α) the equation (6.3) takes the form:

$$\frac{\partial^2 V}{\partial t^2}\frac{1}{V} = \frac{1}{2}\left(\frac{\partial V}{\partial t}\right)^2 \frac{1}{V^2} + \varphi(t), \tag{6.5}$$

i.e.

$$VV_{tt}'' = \frac{1}{2}(V_t')^2 + V^2\varphi(t), t \geq 0 \tag{6.6}$$

$V(0, \alpha) = \frac{\partial x}{\partial \alpha}(0, \alpha) = 1$ (see: $x(0, \alpha) = \alpha$ in (6.4)),

$$\frac{\partial V}{\partial t}(0, \alpha) = \frac{\partial \tilde{u}}{\partial \alpha}(0, \alpha) = F'(\alpha).$$

In fact, (6.3), (6.4) imply that $u(0, x(0, \alpha)) = F(x(0, \alpha)) \Rightarrow \tilde{u}(0, \alpha) = F(\alpha)$.

Fortunately, we can solve the Cauchy problem for (6.6) supposing that $V > 0$ for $t \geq 0$. The change $V = p^2$ gives us $V_t = 2pp'$, $V_{tt} = 2(p')^2 + 2pp'' \Rightarrow$ (if we take $p = \sqrt{V}$),

$$p'' - \frac{\varphi(t)}{2}p = 0 \tag{6.7}$$

$$p(0, \alpha) = 1$$

$$p'(0, \alpha) = F'(\alpha)/2.$$

To simplify our considerations we take an arbitrary function $p_1 \in C^4(t \geq 0)$, $p_1 > 0$ for every $t \geq 0$, $p_1(0) = 1$, $p_1'(0) = 0$ and $p_1 \approx At^\alpha$, $t \to \infty$, where $A = const > 0$, $\alpha > 1/2$. Put $\varphi(t) = \frac{2p_1''}{p_1} \in C^2(t \geq 0) \Rightarrow \varphi \approx \frac{2\alpha(\alpha-1)}{t^2}$, $t \to \infty$. According to the Liouville's formula the function $p_2(t) = p_1(t) \int_0^t \frac{ds}{p_1^2(s)} \in C^4(t \geq 0)$ satisfies the second order linear ODE (6.7) and $p_2(0) = 0$, $p_2'(0) = 1$. Evidently, $p_2(t) > 0$ for $t > 0$, $\int_0^\infty \frac{ds}{p_1^2(s)} = B = const > 0$, i.e. $p_2(t) \approx ABt^\alpha$, $t \to \infty$. Moreover, if we define $g(t) = \frac{p_1}{p_2} = (\int_0^t \frac{ds}{p_1^2(s)})^{-1}$, then $g(t) > 0$ for $t > 0$, $g(t) \in C^4(t > 0)$,

$g'(t) = \frac{p_1'p_2 - p_1p_2'}{p_2^2} = -\frac{W(p_1,p_2)(t)}{p_2^2}$, $W = \begin{vmatrix} p_1 & p_2 \\ p_1' & p_2' \end{vmatrix}$ being the Wronskian of the solutions p_1, p_2 of (6.7). It is well known that $W(t) = W(t_0) = W(0) = 1 \Rightarrow g'(t) = -\frac{1}{p_2^2} < 0$ for $t > 0$. In other words, the function $g(t)$ is strictly monotonically decreasing, $g(t) \sim \frac{1}{t}$, $t \to +0$, $\lim_{t\to\infty} g(t) = \frac{1}{B} > 0$. Moreover, p_1, p_2 are linearly independent functions.

The unique smooth solution of the Cauchy problem for (6.7) is given by:

$$p = p_1(t) + p_2(t)\frac{F'(\alpha)}{2}.$$

Consequently,

$$V = \left(p_1 + \frac{1}{2}F'(\alpha)p_2\right)^2. \tag{6.8}$$

Therefore, the equality $\frac{\partial \tilde{u}}{\partial \alpha} = \frac{\partial V}{\partial t}$ enables us to conclude that

$$\frac{\partial \tilde{u}}{\partial \alpha} = 2p_1\frac{dp_1}{dt} + F'(\alpha)\left(\frac{dp_1}{dt}p_2 + p_1\frac{dp_2}{dt}\right) + p_2\frac{dp_2}{dt}\frac{(F')^2}{2}(\alpha),$$

$\tilde{u}(0, \alpha) = F(\alpha)$, i.e.

$$\tilde{u}(t, \alpha) = Q(t) + 2\alpha p_1 \frac{dp_1}{dt} + F(\alpha) \left(p_2 \frac{dp_1}{dt} + p_1 \frac{dp_2}{dt} \right) \qquad (6.9)$$

$$+ G(\alpha) p_2 \frac{dp_2}{dt} - F(0) \left(p_2 \frac{dp_1}{dt} + p_1 \frac{dp_2}{dt} \right) - G(0) p_2 \frac{dp_2}{dt},$$

where $Q(t) = \tilde{u}(t, 0)$, $G'(\alpha) = \frac{(F')^2(\alpha)}{2}$, i.e. we can take $G(\alpha) = \frac{1}{2} \int_{+\infty}^{\alpha} (F')^2(\xi) d\xi$, $F' \in L_2(\mathbf{R}^1)$. Put $H'(t) = Q(t) - F(0)(p_2 \frac{dp_1}{dt} + p_1 \frac{dp_2}{dt}) - G(0) p_2 \frac{dp_2}{dt}$, $H(0) = 0$. Thus,

$$\tilde{u}(t, \alpha) = H'(t) + 2\alpha p_1 \frac{dp_1}{dt} + F(\alpha) \left(p_2 \frac{dp_1}{dt} + p_1 \frac{dp_2}{dt} \right) + G(\alpha) p_2 \frac{dp_2}{dt}. \quad (6.10)$$

From $\tilde{u}(0, \alpha) = F(\alpha)$ we obtain that $H'(0) = 0$, i.e.

$$\tilde{u}(t, \alpha) = 2\alpha p_1 \frac{dp_1}{dt} + F(\alpha) \left(p_1 \frac{dp_2}{dt} + p_2 \frac{dp_1}{dt} \right) + G(\alpha) p_2 \frac{dp_2}{dt} + H'(t), \quad (6.11)$$

$H'(0) = H(0) = 0$, $G' = \frac{1}{2}(F')^2$.

According to (6.4) $\frac{\partial x}{\partial t} = \tilde{u}(t, \alpha)$, $x(0, \alpha) = \alpha \Rightarrow$

$$x(t, \alpha) = \alpha + \int_0^t \tilde{u}(s, \alpha) ds = \alpha + \alpha(p_1^2(t) - 1) \qquad (6.12)$$

$$+ F(\alpha)(p_1(t)p_2(t) - p_1(0)p_2(0)) + \frac{G(\alpha)}{2}(p_2^2(t) - p_2^2(0)) + H(t)$$

$$= \alpha p_1^2(t) + F(\alpha) p_1(t) p_2(t) + \frac{1}{2} G(\alpha) p_2^2(t) + H(t).$$

Fix $t \geq 0 \Rightarrow p_1^2(t) > 0$ and use the fact that F, G are bounded.

Then from (6.12) we have that the mapping $x : \mathbf{R}_\alpha^1 \to \mathbf{R}_x^1$ is onto. This way we come to the following Theorem.

Theorem 6.1. *Every C^3 smooth solution of (6.2) is given implicitly by*

$$u = \alpha \frac{d}{dt} p_1^2 + F(\alpha) \frac{d}{dt}(p_1 p_2) + G(\alpha) \frac{1}{2} \frac{d}{dt} p_2^2 + H'(t) \qquad (6.13)$$

$$= \frac{d}{dt} (\alpha p_1^2 + F(\alpha) p_1 p_2 + \frac{1}{2} G(\alpha) p_2^2 + H(t)),$$

$0 = H(0) = H'(0)$,

$$x = \alpha p_1^2(t) + F(\alpha) p_1 p_2 + \frac{1}{2} G(\alpha) p_2^2 + H(t), \qquad (6.14)$$

where $G'(\alpha) = \frac{1}{2}(F')^2(\alpha)$ *and* p_1, p_2 *are linearly independent solutions of* (6.7): $p'' = \frac{\varphi(t)}{2}p$, $p_1(0) = 1$, $p_1'(0) = 0$, $p_2(0) = 0$, $p_2'(0) = 1$.

To complete the things we must find

$$\frac{\partial x}{\partial \alpha} = p_1^2(t) + F'(\alpha)p_1 p_2 + \frac{1}{2}G'(\alpha)p_2^2$$

$$= \left(p_1 + \frac{F'(\alpha)}{2}p_2\right)^2 = p_2^2\left(g(t) + \frac{F'(\alpha)}{2}\right)^2 \geq 0, t > 0.$$

Evidently, $\frac{\partial x}{\partial \alpha}|_{t=0} = 1$. So we are interested when $g(t) + \frac{F'(\alpha)}{2} > 0$ for $t > 0$ and $\forall \alpha \in \mathbf{R}^1$.

To simplify the study let $H \equiv 0$ (we are dealing with the existence problem), i.e.

$$u(t, \alpha) = \frac{d}{dt}(\alpha p_1^2(t) + F(\alpha)p_1 p_2 + \frac{1}{2}G(\alpha)p_2^2(t)), \tag{6.15}$$

$$x(t, \alpha) = \alpha p_1^2(t) + F(\alpha)p_1 p_2 + \frac{1}{2}G(\alpha)p_2^2.$$

We are looking for $t^* > 0$ and such that $\frac{\partial x}{\partial \alpha} = p_2^2(t)(g(t) + \frac{F'(\alpha)}{2})^2 > 0$, $\forall \alpha \in \mathbf{R}^1$ and for each t: $t^* > t > 0$, as $\frac{\partial x}{\partial \alpha}|_{t=0} = 1$. Then according to the inverse mapping theorem there exists a unique smooth mapping $\alpha = \alpha(t, x)$, $0 \leq t < t^*$, t being a parameter. In fact, consider the mapping $x(t, \alpha) : \mathbf{R}_\alpha^1 \to \mathbf{R}_x^1$ smoothly depending on the parameter $t \geq 0$. As we mentioned (see (6.15)) for each fixed $t \geq 0$ it is onto. If $\frac{\partial x}{\partial \alpha} > 0$ for $\forall \alpha \in \mathbf{R}^1$ and for $0 \leq t < t^*$ then there exists the inverse mapping $\alpha = \alpha(t, x) \in C^2([0, t^*) \times \mathbf{R}_x^1)$. Then $u = u(t, \alpha(t, x))$ given by (6.15) is a solution of (6.2); $\alpha \to \pm\infty \iff x(t, \alpha) \to \pm\infty$.

Remark 6.1. Compute $\frac{\partial u}{\partial x} = \frac{\partial \tilde{u}}{\partial \alpha}\frac{\partial \alpha}{\partial x} = \frac{\frac{\partial V}{\partial t}}{V}$ ($V > 0$ everywhere). Having in mind that $p_2(0) = 0$, $p_2 > 0$ for $t > 0$ and $\lim_{t \to +0}(g(t) + \frac{F'(\alpha)}{2}) = +\infty$ we conclude that we must find the maximal interval $I = [0, t^*)$ such that $g(t) + \frac{F'(\alpha)}{2} > 0$, $\forall t \in [0, t^*)$ and for each $\alpha \in \mathbf{R}^1$. Thus, for $t > 0$

$$\frac{\partial u}{\partial x} = \frac{\frac{\partial V}{\partial t}}{V} = \frac{\partial}{\partial t}\ln V = 2\frac{\partial}{\partial t}\ln p_2 + 2\frac{\partial}{\partial t}\ln\left(g + \frac{F'}{2}\right) = 2\frac{p_2'}{p_2} + 2\frac{g'(t)}{g(t) + \frac{F'(\alpha)}{2}}.$$

We know that $2\frac{p_2'}{p_2} \in C^2(t > 0)$, $\frac{g'(t)}{g(t) + \frac{F'(\alpha)}{2}} \in C^2([0, t^*) \times \mathbf{R}_\alpha^1)$ and $g'(t) < 0$ for each $t > 0$ and therefore $\frac{\partial u}{\partial x} = -\infty$ if $g(t^*) + \frac{F(\alpha^*)}{2} = 0$ for some α^*.

Assume now that for some $\tilde{\alpha}: F'(\tilde{\alpha}) < 0$ and $0 < T = sup(\frac{-F'(\alpha)}{2})$. For the sake of simplicity let $sup(\frac{-F'(\alpha)}{2}) = -\frac{F'}{2}(\alpha_0)$ and α_0 is unique.

On the other hand, $g(t) > \frac{1}{B}$ for each $t > 0$, $g(+0) = +\infty$, $g(+\infty) = \frac{1}{B}$. In the case $\frac{1}{B} \geq -\frac{F'}{2}(\alpha_0) = T$ we have that $\forall t \geq 0$, $\forall \alpha$: $g(t) > \frac{1}{B} \geq -\frac{F'(\alpha_0)}{2} \geq -\frac{F'(\alpha)}{2} \Rightarrow g(t) + \frac{F'(\alpha)}{2} > 0$, while in the case $\frac{1}{B} < T$ there exists a unique point $t^* > 0$ and such that $g(t^*) = T$, $0 < t < t^* \Rightarrow g(t) > T$. Thus, $g(t) > T \geq -\frac{F'(\alpha)}{2}$ for each $0 \leq t < t^*$, $\forall \alpha$; $g(t^*) + \frac{F'(\alpha_0)}{2} = 0$.

Theorem 6.2. *(i) The Cauchy problem for (6.2) possesses a global in time $t \geq 0$ classical solution u if either $F'(\alpha) \geq 0$, $\forall \alpha$ or if $inf_{t\geq0} \frac{p_1}{p_2} \geq sup\frac{(-F')}{2}$.*

(ii) Conversely, if $inf_{t\geq0} \frac{p_1}{p_2} < sup\frac{(-F'(\alpha))}{2} = T$ there exists a classical solution u of (6.2), $u \in C^3([0, t^*) \times \mathbf{R}_x^1)$, where $\frac{p_1(t^*)}{p_2(t^*)} = T$ and such that $u_x \to -\infty$ as $t \nearrow t^*$, $x_{\alpha\to\alpha_0} \to x_0$, $x_0 = x(t^*, \alpha_0)$.

In other words, in the case (ii) u_x blows up as $t \to t^*$ and the life span of u is t^*.

Remark 6.2. The Cauchy problem (6.1) has a unique classical solution under the additional assumption $u \to 0$ for $x \to \infty$ and for each fixed $t \geq 0$. This is a boundary condition, of course. Thus, $F(x) \to 0$ for $x \to +\infty$, $\varphi(t) \equiv 0$. One can easily see that $p_1 = 1$, $p_2(t) = t$, $g = \frac{p_1}{p_2}$, $inf_{t\geq0}g = 0$. Consequently, $u = F(\alpha) + tG(\alpha) + H'(t)$, $H(0) = H'(0) = 0$. The condition $G(+\infty) = 0$ implies that $H'(t) \equiv 0$, $\forall t \geq 0 \Rightarrow H(t) = 0$ because $\alpha \to +\infty \Rightarrow x(t, \alpha) \to \infty$. Therefore, u satisfies the relations: $u = F(\alpha) + tG(\alpha)$, $x = \alpha + tF(\alpha) + \frac{t^2}{2}G(\alpha)$ etc. We point out that in this special case a global in time solution u exists if and only if $F'(\alpha) \geq 0$, $\forall \alpha \in \mathbf{R}^1$.

If $F' \geq 0$ everywhere $\Rightarrow \frac{\partial x}{\partial \alpha} > 0$ and for each fixed $\tilde{t}$ we have that $x \to +\infty \Longleftrightarrow \alpha \to +\infty$.

Assume now that $\frac{d}{dt}p_1^2(\tilde{t}) \neq 0$. Then (6.13) implies that $|u(\tilde{t}, \alpha)| \to \infty$ for $x \to \infty$ $(\alpha = \alpha(\tilde{t}, x))$.

6.2 Weak solutions of the Hunter-Saxton equation

1. We can find the formula for the classical solution u of the Cauchy problem for the reduced first order Hunter-Saxton equation $u_t + uu_x = g(t, x)$, $u \to 0$ for $x \to \infty$ avoiding the standard changes of the variables proposed above

(see (6.3), (6.5), (6.6) with $\varphi \equiv 0$, etc.). Moreover, the new approach works in the case of constructing of admissible weak solution, i.e. generalized solution of (6.1). We observe that there is an associated conservation law with the Hunter – Saxton equation, namely: $(u_x^2)_t + (uu_x^2)_x = 0$. In other words (6.1) implies that $(u_x^2)_t + (uu_x^2)_x = 0$, $t \geq 0$. We omit the trivial calculation. As we know (see (6.4)) the characteristic $\mathcal{L} : x = x(t, \alpha)$, shortly $x = \gamma(t)$, satisfies the ODE $\frac{d\gamma}{dt} = u(t, \gamma(t))$, $\gamma(0) = \alpha \in \mathbf{R}^1 \Rightarrow$ $\frac{d}{dt} u(t, \gamma(t)) = u_t + \gamma' u_x = \frac{1}{2} \int_{-\infty}^{\gamma(t)} u_x^2(t, y) dy = g(t, \gamma(t))$ as by definition $g(t, x) = \frac{1}{2} \int_{-\infty}^{x} u_x^2(t, y) dy$. Suppose now that $F'(x) \geq 0$. Let $\mathcal{L}_1 : x = \gamma(t)$, $\mathcal{L}_2 : x = \delta(t)$ be two characteristics globally defined for $t \geq 0$ and starting from the points $\delta(0) \leq \gamma(0)$.

Integrating the associated conservation law from $\delta(t)$ to $\gamma(t)$ with respect to the variable x we get:

$$\int_{\delta(t)}^{\gamma(t)} (u_x^2(t, x))_t dx + u(t, \gamma(t)) u_x^2(t, \gamma(t)) - u(t, \delta(t)) u_x^2(t, \delta(t)) = 0,$$

i.e.

$$\frac{\partial}{\partial t} \int_{\delta(t)}^{\gamma(t)} u_x^2(t, x) dx = \int_{\delta(t)}^{\gamma(t)} (u_x^2(t, x))_t dx + \gamma' u_x^2(t, \gamma) - \delta' u_x^2(t, \delta)$$

$$= \int_{\delta(t)}^{\gamma(t)} (u_x^2(t, x))_t dx + u(t, \gamma) u_x^2(t, \gamma) - u(t, \delta) u_x^2(t, \delta) = 0$$

$$\Rightarrow \int_{\delta(t)}^{\gamma(t)} u_x^2(t, x) dx = const = \int_{\delta(0)}^{\gamma(0)} u_x^2(0, x) dx = \int_{\delta(0)}^{\gamma(0)} (F')^2(x) dx < \infty.$$

(6.16)

Suppose that $\gamma(0) = \delta(0)$. According to (6.16) we obtain that $\gamma(t) = \delta(t)$, $\forall t \geq 0$ and consequently we have global uniqueness result for the characteristics. This result is well known as $u \in C^2 (t \geq 0)$ and $\mathcal{L}$ satisfies (6.4). Moreover, if $\delta(0) < \gamma(0)$ then $\delta(t) < \gamma(t)$, $\forall t \geq 0$, i.e. if the characteristic curve $\mathcal{L}_1$ starts from the point $\gamma(0) > \delta(0)$ then $\mathcal{L}_1$ is located above $\mathcal{L}_2$ for each $t > 0$. In other words, $\mathcal{L}_1$ and $\mathcal{L}_2$ are not crossing smooth curves. Suppose that $\gamma(t) \to +\infty$ for $t \to \infty$ and $\delta(t) \to -\infty$ for $t \to +\infty$. Then we claim that

$$\int_{-\infty}^{\infty} u_x^2(t, x) dx = const.$$

(6.17)

To prove the conservation law (6.17) we consider the identity

$$\frac{\partial}{\partial t} \int_{-\infty}^{\gamma(t)} u_x^2(t, x) dx = \int_{-\infty}^{\gamma(t)} \frac{\partial}{\partial t} (u_x^2(t, x)) dx + \gamma' u_x^2(t, \gamma); \quad \gamma' = u(t, \gamma).$$

To simplify our investigation we could assume that $\int_{-\infty}^{\infty} u_x^2(t,x)dx < \infty \Rightarrow u_x^2(t,x) \to 0$ for $x \to \pm\infty$.

On the other hand,

$$\int_{-\infty}^{\gamma(t)} (u_x^2)_t(t,x)dx + u(t,\gamma)u_x^2(t,\gamma) - \lim_{x\to-\infty} u(t,x)u_x^2(t,x) = 0.$$

Under the additional assumption $\lim_{x\to-\infty} u(t,x)u_x^2(t,x) = 0$ we obtain that

$$\frac{\partial}{\partial t}\int_{-\infty}^{\gamma(t)} u_x^2(t,x)dx = 0, \quad \forall t \geq 0. \tag{6.18}$$

Thus

$$\frac{1}{2}\int_{-\infty}^{\gamma(t)} u_x^2(t,x)dx = \frac{1}{2}\int_{-\infty}^{\gamma(0)} u_x^2(0,x)dx = \frac{1}{2}\int_{-\infty}^{\alpha} (F')^2(x)dx = const, \tag{6.19}$$

i.e.

$$g(t,\gamma(t)) = g(0,\alpha), \quad \forall t \geq 0; \ \gamma(0) = \alpha \in \mathbf{R}^1. \tag{6.20}$$

This way we conclude that:

$$\frac{d\gamma}{dt} = u(t,\gamma(t)), \gamma(0) = \alpha, \frac{d}{dt}u(t,\gamma(t)) = g(t,\gamma(t)) = g(0,\alpha), \tag{6.21}$$

$u|_{t=0} = F(\alpha)$.

After a simple integration we have:

$$u(t,\gamma(t)) = F(\alpha) + \frac{t}{2}\int_{-\infty}^{\alpha} (F')^2(x)dx, \tag{6.22}$$

$$\gamma(t) = \alpha + tF(\alpha) + \frac{t^2}{4}\int_{-\infty}^{\alpha} (F')^2(x)dx. \tag{6.23}$$

Our next step is to invert the transform $(t,\alpha) \to (t,x)$ defined by

$$x = \alpha + tF(\alpha) + \frac{t^2}{4}\int_{-\infty}^{\alpha} (F')^2(x)dx. \tag{6.24}$$

We observe that for each $\alpha \in \mathbf{R}^1$ fixed the characteristic (6.23) is a globally defined parabola in $\mathbf{R}_{t,x}^2$, $t \geq 0$, α being a parameter. Evidently, $\frac{\partial x}{\partial \alpha} = 1 + tF'(\alpha) + \frac{t^2}{4}(F')^2(\alpha) = (1 + \frac{t}{2}F'(\alpha))^2 > 0$ for $F' \geq 0$ and $\forall t \geq 0$. The assumptions $F \in L^\infty(\mathbf{R}^1)$, $\int_{-\infty}^{\infty}(F')^2(x)dx < \infty$ imply that if $\alpha \to \pm\infty$ in (6.24) then for each fixed $t \geq 0$ we have that $x \to \pm\infty$. The implicit function theorem with parameter t gives us that there exists an

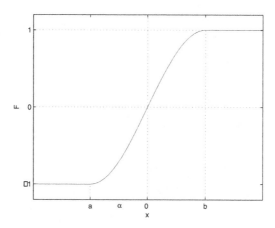

Fig. 6.1

inverse function $\alpha = \alpha(t, x) \in C^2(t \geq 0)$ of $x = x(t, \alpha)$. Putting in (6.22) $\alpha = \alpha(t, x)$ instead of $x = x(t, \alpha) = \gamma(t)$ we obtain the solution

$$u(t, x) = F(\alpha(t, x)) + \frac{t}{2} \int_{-\infty}^{\alpha(t,x)} (F')^2(\lambda) d\lambda \qquad (6.25)$$

of the Cauchy problem (6.1), $u \in C^2(t \geq 0)$.

We shall illustrate the above made study by a simple example.

Example 1. Define now the function $F(x) \in C^2(\mathbf{R}^2)$, $-1 \leq F(x) \leq 1$, $F \equiv -1$ for $x \leq a < 0$, $F(x) \equiv 1$ for $x \geq b > 0$, $F'(x) > 0$ for $a < x < b$, $F(0) = 0$ (see Fig. 6.1).

The characteristics (6.23) of the corresponding Hunter-Saxton equation $u_t + u u_x = g(t, x)$ are of three different types.

I. $\alpha \leq a$. Then $x = \alpha - t$ (parallel straight lines), $t \geq 0$.

II. $a < \alpha < b$. Then $\mathcal{L} : x = \alpha + tF(\alpha) + \frac{t^2}{4} \int_a^\alpha (F')^2(\lambda) d\lambda$. Certainly, $\mathcal{L}$ is a parabola in $\mathbf{R}^2_{tx}$ starting for $t = 0$ from α. $\mathcal{L}$ has an axis parallel to the $\overrightarrow{Ox}$ axis and a vertex at the point $A(\alpha) = (t(\alpha), x(\alpha))$, where $t(\alpha) = -\frac{2F(\alpha)}{\int_a^\alpha (F')^2(\lambda)d\lambda}$, $x(\alpha) = \alpha + t(\alpha)F(\alpha) + \frac{t^2(\alpha)}{4} \int_a^\alpha (F')^2(\lambda)d\lambda = \alpha - \frac{F^2(\alpha)}{\int_a^\alpha (F')^2(\lambda)d\lambda}$. Evidently, $\alpha \to a + 0 \Rightarrow t(\alpha) \to +\infty$, $x(\alpha) \to -\infty$. $\alpha \to b - 0 \Rightarrow t(\alpha) \to A = \frac{-2}{\int_a^b (F')^2(\lambda)d\lambda} = const < 0$, $a < \alpha < 0 \Rightarrow t(\alpha) > 0$, $t(0) = 0$, $x(0) = 0$; $0 < \alpha \leq b \Rightarrow t(\alpha) < 0$.

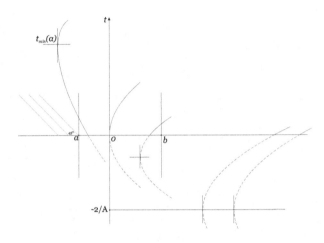

Fig. 6.2

III. $\alpha \geq b \Rightarrow x = \alpha + t + \frac{t^2}{4}B$, $B = \int_a^b (F')^2(\lambda)d\lambda$. In this case $t(\alpha) = -\frac{2}{B} = A < 0$, i.e. the parabolas of the third type have the same axis – the ray passing through the point $(-\frac{2}{B}, 0)$ and parallel to $\overrightarrow{Ox}$.

The geometrical interpretation of these characteristics in $t \geq 0$ which are not crossing each other is given on Fig. 6.2.

2. We propose below a short survey on some results concerning the generalized solutions of the Cauchy problem for the Hunter-Saxton equation realized as a balance law:

$$u_t + \left(\frac{u^2}{2}\right)_x = g(t, x), \quad t \geq 0 \tag{6.26}$$

$$u(0, x) = u_0(x), \quad u_x(0, x) = u_{0x}, \quad -\infty < x < \infty, \tag{6.27}$$

where the source term $g(t, x) = \frac{1}{2}\int_{-\infty}^x w^2(t, y)dy$, $w(t, x) = u_x(t, x)$, $u_0 \in L^\infty(\mathbf{R}^1)$ is a bounded absolutely continuous function whose derivative (a.e.) $u_{0x} \in L_2(\mathbf{R})$.

Definition 6.1 (6). *We shall say that the bounded continuous function $u(t, x)$ on the strip $[0, T) \times \mathbf{R}_x^1$ is a weak solution of Hunter-Saxton equation (6.26) if: a) for every fixed $t \in [0, T)$ the function $u(t, .) : \mathbf{R}_x^1 \to \mathbf{R}^1$ is absolutely continuous, its derivative $u_x = w$ is such that $w(t, .) \in L^\infty([0, T) : L^2(\mathbf{R}_x^1))$, and b) (6.26) holds in the sense of distributions.*

We point out that the absolutely continuous functions are differentiable a.e. Moreover, (6.26) is satisfied in generalized sense (D') if

$$\int_0^T \int_{-\infty}^{\infty} \left(u\varphi_t + \frac{1}{2}u^2\varphi_x + g\varphi \right) dx dt = 0$$

for each test function $\varphi(t, x)$.

Having in mind that u is continuous we conclude that $t \longmapsto w(t,.)$ as a map from $[0, T)$ to $L^2(\mathbf{R}_x^1)$ is at least weakly continuous. In (6.27) we can write $u(0, x) = u_0(x)$, $w(0, x) = w_0(x) = u_{0x}$ instead of $u_x(0, x) = u_{0x}$.

It is interesting to mention that in general (6.26), (6.27) possesses multiple solutions. We did this observation in studying the classical Cauchy problem (6.2) ((6.3)) too. For example, both $u \equiv 0$ and

$$\begin{cases} 0, & t > 0, \quad x \le 0 \\ \frac{2x}{t}, & t > 0, \quad 0 < x \le \frac{t^2}{4} \\ \frac{t}{2}, & t > 0, \quad \frac{t^2}{4} < x \end{cases}$$

are weak solutions of (6.26), (6.27) with initial data $u_0 = 0$, $w_0 = 0$.

In order to have a uniqueness result we give the following definition.

Definition 6.2. [40] The weak solution u of the Cauchy problem (6.26), (6.27) is called admissible if: c) w is uniformly bounded from above on any compact subset of the strip $(0, T) \times \mathbf{R}_x^1$ and d) $w(t, .) \to w_0(.)$ strongly in $L^2(\mathbf{R}_x^1)$ for $t \searrow 0$.

This is the main result concerning the Cauchy problem (6.26), (6.27).

Theorem 6.3. [40] *For any nondecreasing bounded absolutely continuous initial data u_0 with $u_{0x} = w_0 \in L^2(\mathbf{R}_x^1)$ there exists a unique admissible solution u of (6.26), (6.27) on $[0, \infty) \times \mathbf{R}_x^1$. Moreover, $u(t, x)$ is locally Lipschitz on the strip $[0, \infty) \times \mathbf{R}_x^1$ and satisfies the equation (6.26) almost everywhere.*

Certainly, u_0 nondecreasing implies that $w_0 = u_{0x} \ge 0$ a.e. The detailed proof of Theorem 6.3 could be found in [40] (see also [38] and [18]).

3. We shall complete this Chapter with a short remark on the propagation of jump type singularities of the solutions of the second order nonlinear PDE (6.1). To begin with, we shall remind of the reader the definition of the characteristics of the quasilinear PDE ([36])

$$ar + bs + c\tilde{t} + d = 0, \tag{6.28}$$

where $r = u_{tt}$, $s = u_{tx}$, $\tilde{t} = u_{xx}$ and the coefficients a, b, c, d depend smoothly on (t, x, u, p, q), $p = u_t$, $q = u_x$.

We shall say that the smooth curve

$$\mathcal{L}: \begin{vmatrix} t = t(\lambda) \\ x = x(\lambda) \end{vmatrix}, \ (t')^2 + (x')^2 > 0$$

is a characteristic for (6.28) if it satisfies the ODE $a(x')^2 - bx't' + c(t')^2 = 0$.

Certainly, in a, b, c, d we write $u(t(\lambda), x(\lambda))$, $p(t(\lambda), x(\lambda))$ and $q(t(\lambda), x(\lambda))$. In our case of the equation (6.1) we have: $2u_{tx} + u_x^2 + 2uu_{xx} = 0$, i.e. $a = 0$, $b = 2$, $c = 2u$, $d = u_x^2$. Thus, the characteristics $\mathcal{L}$ of (6.1) are given by $t'(x' - ut') = 0$. Supposing that $t' \neq 0 \Rightarrow x = x(t)$ we obtain $x' = ut' \Leftrightarrow \frac{dx}{dt} = u(t, x(t))$. This curve coincides with (6.4), evidently.

As (6.1) is second order nonlinear PDE it is quite reasonable in studying propagation of singularities to suppose that $u \in C^1(\Omega)$ but its second derivatives have finite jumps along some smooth curve Γ. The problem is whether Γ coincides (at least locally) with some arc of the characteristic $\mathcal{L}$.

Proposition 6.1. [36] *Suppose that the smooth non-degenerate curve* Γ: $\begin{vmatrix} t = t(\lambda) \\ x = x(\lambda) \end{vmatrix}$ *divides the upper half plane* $\Omega = \{t > 0\}$ *into two parts:* Ω_+ *and* Ω_-. *Let* $u \in C^1(\overline{\Omega}) \cap C^2(\overline{\Omega}_-) \cap C^2(\overline{\Omega}_+)$ *satisfy (6.1) in* Ω_+ *and* Ω_- *and*

$$\partial_\tau(\nabla u_-) = \partial_\tau(\nabla u_+) \ \text{ on } \ \Gamma \tag{6.29}$$

$$\partial_n(\nabla u_-) \neq \partial_n(\nabla u_-) \ \text{ on } \ \Gamma. \tag{6.30}$$

Then Γ *is an arc of a characteristic of (6.1). As usual,* ∂_τ *stands for the directional derivative of* Γ, *while* ∂_n *stands for the normal derivative to* Γ; $u_{pm} = u|_{\Omega_\pm}$.

Proof. Fix the point $P \in \Gamma$. Then the tangential vector $\vec{\tau}$ to Γ is $\vec{\tau} = (t'(\lambda), x'(\lambda)) \neq 0$ and the normal vector $\vec{n}$ is $\vec{n} = (x'(\lambda), -t'(\lambda))$. As we know $u_\pm$ satisfies (6.1) in $\Omega_\pm$ and by continuity in $\overline{\Omega}_\pm$. Consequently,

$$au_{-tt} + bu_{-tx} + cu_{-xx} + d = 0 \ \text{ in } \ \overline{\Omega}_-,$$

$$au_{+tt} + bu_{+tx} + cu_{+xx} + d = 0 \ \text{ in } \ \overline{\Omega}_+,$$

i.e.

$$a[u_{tt}]_\Gamma + b[u_{tx}]_\Gamma + c[u_{xx}]_\Gamma = 0, \tag{6.31}$$

where $[v]_\Gamma = v_+ - v_-$ stands for the finite jump of some function v on Γ.

By definition

$$\frac{\partial u}{\partial \tau} = (\nabla_{t,x} u(t(\lambda), x(\lambda)), \vec{\tau}) = u_t t' + u_x x',$$

$$\frac{\partial u}{\partial n} = (\nabla_{t,x} u(t(\lambda), x(\lambda)), \vec{n}) = x' u_t - t' u_x.$$

Consider now $\frac{\partial u_{+x}}{\partial \tau} = t' u_{+tx} + x' u_{+xx}$, $\frac{\partial u_{-x}}{\partial \tau} = t' u_{-tx} + x' u_{-xx}$ and apply (6.29). Thus,

$$\begin{vmatrix} t'[u_{tx}] + x'[u_{xx}] = 0 \\ t'[u_{tt}] + x'[u_{tx}] = 0. \end{vmatrix} \tag{6.32}$$

The linear system (6.32) possesses a nontrivial solution (t', x'), as $(t')^2 + (x')^2 > 0$. Therefore

$$[u_{tx}]_\Gamma^2 = [u_{xx}]_\Gamma [u_{tt}]_\Gamma. \tag{6.33}$$

Certainly, (6.30) implies that $[u_{xx}]_\Gamma^2 + [u_{tx}]_\Gamma^2 + [u_{tt}]_\Gamma^2 > 0$.

We must consider two different cases, namely

1) $[u_{tx}] \neq 0 (\Rightarrow [u_{xx}], [u_{tt}] \neq 0)$ and $[u_{tx}] = 0 (\Rightarrow$ either $[u_{xx}] = 0$ or $[u_{tt}] = 0)$.

To fix the ideas we shall concentrate on case 1) only. Then (6.32) implies that $t' = k[u_{tx}]_\Gamma$; $k = k(\lambda) \neq 0$, $x' = -k[u_{tt}]_\Gamma$. Combining (6.31) and (6.33) we get: $ax' - bt' + c\frac{(t')^2}{x'} = 0$, i.e. $a(x')^2 - bx't' + c(t')^2 = 0$. Evidently, $\frac{dx}{dt} = -\frac{[u_{tx}]}{[u_{xx}]} = u$. Proposition 6.1 is proved.

In other words the finite jumps of the second derivative (the Hessian of u) propagate along the characteristics of the equation (6.28).

Chapter 7

Integrable multicomponent generalizations of the Camassa-Holm equation

7.1 Construction of peakon-type solutions of the Hunter-Saxton equation

1. The most popular multicomponent generalization of the Camassa-Holm equation is the following one:

$$\left| \begin{array}{l} m_t + 2u_x m + um_x + \sigma\rho\rho_x = 0 \\ \rho_t + (u\rho)_x = 0, \end{array} \right. \tag{7.1}$$

where $m = \sigma_1 u - u_{xx}$, $\sigma = \pm 1$, $\sigma_1 = 1$ or in the "short wave" limit $\sigma_1 = 0$. To fix the ideas and to simplify the things we shall deal with peakon solutions of the "short wave" limit equation $\sigma_1 = 0$ and under the assumption $\rho = 0$. This is the Hunter-Saxton equation studied in Chapter VI:

$$m_t + 2u_x m + um_x = 0, m = -u_{xx}, \tag{7.2}$$

i.e.

$$u_{xxt} + 2u_x u_{xx} + uu_{xxx} = 0; u = u(x,t) \in D'. \tag{7.3}$$

As we discussed in the previous Chapter the existence of weak solutions of (7.3), we shall concentrate here on the construction of peakon solutions of (7.3) (see [58] and [33]).

To generalize our considerations and to compare the behavior of the waves for different coefficients we will study peakon solutions of

$$u_{xxt} + 2su_x u_{xx} + uu_{xxx} = 0, \tag{7.4}$$

i.e.

$$m_t - s\frac{\partial}{\partial x}(u_x^2) + um_x = 0, \tag{7.5}$$

where $s = 1$ or $s = 2$.

We are looking for peakon solutions of (7.5) having the form (see [33]):

$$m(x,t) = \sum_{k=1}^{N} m_k(t)\delta(x - x_k(t)), N \geq 2, \tag{7.6}$$

$$u(x,t) = -\frac{1}{2}\sum_{k=1}^{N} m_k(t)|x - x_k(t)|,$$

$t > const \geq 0$ and under the condition:

For each $t > const \geq 0, t$ fixed : $u(x,t) \to l(t)$ for $x \to \infty$, (7.7)

where $l(t)$ is a continuous function depending on t.

We shall see below that in Schwartz distribution sense $u_{xx}(x,t) = -m(x,t)$. Certainly, $u, m \in D'(\mathbf{R}_x^1 \times \{t > c\})$, $c = const$.

Proposition 7.1. *Let $\psi \in C^1(a > const)$. Define the distribution $\delta(x - \psi(a)) \in D'(\mathbf{R}_x^1 \times \{a > const\})$ by the formula*

$$< \delta(x - \psi(a)), \varphi(x,a) >= \int_{-\infty}^{\infty} \varphi(\psi(a),a)da, \forall \varphi \in C_0^0(\mathbf{R}_x^1 \times \{a > const\}).$$

Then

$$\frac{\partial}{\partial a}(\delta(x - \psi(a))) = -\psi'(a)\delta'(x - \psi(a)). \tag{7.8}$$

The derivatives are taken in D'.

Proof: According to the definition of a derivative of Schwartz distribution we have

$$< \frac{\partial}{\partial a}(\delta(x - \psi(a))), \varphi(a) >= - < \delta(x - \psi(a)), \frac{\partial\varphi}{\partial a} > \tag{7.9}$$

$$= -\int_{-\infty}^{\infty} \frac{\partial\varphi}{\partial a}(\psi(a),a)da = \int_{-\infty}^{\infty} \frac{\partial\varphi}{\partial a}(\psi(a),a)\psi'(a)da - \int_{-\infty}^{\infty} \frac{d}{da}(\varphi(\psi(a),a)da$$

$$=< \psi'(a)\delta(x - \psi(a)), \frac{\partial\varphi}{\partial x}(x,a) >= - < \frac{\partial}{\partial x}(\psi'(a)\delta(x - \psi(a))), \varphi > .$$

Thus (7.8) is proved.

Applying (7.8) to u with $a = t$ we get in D':

$$u_{xx} = -1/2 \sum_{k=1}^{N} m_k(t) \frac{\partial^2}{\partial x^2} |x - x_k(t)|.$$

According to the jump formula in D' (see [52]) and Chapter V we get $\frac{\partial}{\partial x} |x - x_k(t)| = sgn(x - x_k(t))$, $\frac{\partial^2}{dx^2} |x - x_k(t)| = 2\delta(x - x_k(t)) \Rightarrow u_{xx} = -m$ in $D'(\mathbf{R}_x^1 \times \{t > const\})$. Fix $t > const$ and let $x \to +\infty$ in (7.6). So we can assume that $x > max_{1 \leq k \leq N} x_k(t) \Rightarrow u(x,t) = (-1/2 \sum_{k=1}^{N} m_k(t))x + \frac{1}{2} \sum_{k=1}^{N} m_k(t)x_k(t)$. Then (7.7) implies that $\sum_{k=1}^{N} m_k(t) = 0, \forall t > const$ and $l(t) = \frac{1}{2} \sum_{k=1}^{N} m_k(t)x_k(t)$.

Further on we shall assume that

$$x_1(t) < x_2(t) < \ldots < x_N(t), \forall t > const. \tag{7.10}$$

More precisely, we shall impose additional conditions guaranteeing the fulfilment of (7.10). Substituting (7.6) into (7.5) we get (formally!) in D':

$$m_t = \sum_{k=1}^{N} m_k'(t)\delta(x - x_k(t)) - \sum_{k=1}^{N} m_k(t)x_k'(t)\delta'(x - x_k(t)),$$

$u_x = -1/2 \sum_{k=1}^{N} m_k(t)sgn(x - x_k(t)) \Rightarrow$ (formally):

$$m_x = \sum_{k=1}^{N} m_k(t)\delta'(x - x_k(t)) \Rightarrow$$

(formally):

$$um_x = -1/2 \sum_{k=1}^{N} m_k(t)\delta'(x - x_k(t)). \sum_{l=1}^{N} m_l(t)|x - x_l(t)|.$$

We shall apply now the jump formula in $D'(\mathbf{R}_x^1 \times \{t > const\})$ for the distribution u_x^2 having finite jumps on the smooth curves $\Gamma_j : x = x_j(t)$, $t > const$, $1 \leq j \leq N$. Thus

$$\partial_x(u_x^2) = \{\partial_x(u_x^2)\} + \sum_{k=1}^{N} [u_x^2]_{\Gamma_k} \delta(x - x_k(t)).$$

The symbol $\{\partial_x(u_x^2)\}$ stands for the values of the function $\partial_x(u_x^2)$ outside the closed set $\Gamma = \cup_{j=1}^{N} \bar{\Gamma}_j$, where it is smooth:

$$u_x^2 = 1/4 \sum_{k=1}^{N} \sum_{l=1}^{N} sgn(x - x_k)sgn(x - x_l)m_k m_l.$$

Certainly, $sgn(x - x_k)/sgn(x - x_l)/ = \pm 1$ outside Γ and therefore $\frac{\partial}{\partial x}(u_x^2) = 0$ outside Γ.

To find the jump $[u_x^2]_{\Gamma_k}$ we consider the term $1/4m_k \, sgn(x - x_k) \times \sum_{l=1}^{N} m_l \, sgn(x - x_l) = 1/4m_k^2 \, [sgn(x - x_k)]^2 + 1/4m_k \, sgn(x - x_k) \times \sum_{l \neq k} m_l \, sgn(x - x_l)$. Having in mind that $sgn^2(x - x_k) = 1$, the jump of $sgn(x - x_k)$ along Γ_k is 2 and $sgn(x - x_k)$ is continuous with respect to (x, t) near Γ_l for $l \neq k$, we get:

$$\left[1/4m_k \, sgn(x - x_k) \sum_{l=1}^{N} m_l \, sgn(x - x_l) \right]_{\Gamma_k} = \frac{1}{2} m_k \sum_{k \neq l} m_l \, sgn(x_k - x_l)$$

$$= 1/2m_k \sum_{l=1}^{N} m_l \, sgn(x_k - x_l)$$

with the convention $sgn(0) = 0$.

Thus,

$$-s\partial_x(u^2) = -\frac{s}{2} \sum_{k=1}^{N} m_k \sum_{l=1}^{N} m_l \, sgn(x_k - x_l)\delta(x - x_k).$$

Equalizing the coefficients in front of $\delta(x - x_k(t))$ and $\delta'(x - x_k(t))$ in (7.5) (*supp* $\delta(x - x_k(t)) = \{(x, t) : x = x_k(t)\}$) we get:

$$m_k'(t) = \frac{s}{2} m_k \sum_{l=1}^{N} m_l(t) \, sgn(x_k(t) - x_l(t)) \qquad (7.11)$$

$$-m_k(t)x_k'(t) = \frac{1}{2} m_k(t) \sum_{l=1}^{N} m_l(t)|x_k(t) - x_l(t)|.$$

We rewrite (7.11) as

$$m_k' = \frac{s}{2} m_k \sum_{l=1}^{N} m_l(t) \, sgn(x_k(t) - x_l(t)) \qquad (7.12)$$

$$x_k' = -1/2 \sum_{l=1}^{N} m_l(t)|x_k(t) - x_l(t)|,$$

$$\sum_{l=1}^{N} m_l(t) = 0 \Rightarrow m_1 = -\sum_{2}^{N} m_l(sgn(0) = 0).$$

Put $M_k = \sum_{j=1}^{k} m_j$, $k = 1, \ldots, N - 1$; $M_N \equiv 0$ according to (7.7).

Evidently, $k = 1 \Rightarrow m_1' = -\frac{s}{2} m_1 \sum_{l=2}^{N} m_l$ (see (7.10)), i.e. $m_1' = \frac{s}{2}m_1^2 \iff M_1' = \frac{s}{2}M_1^2$.

Let $N - 1 \geq k \geq 2$. Then from (7.12) we have:

$$M_k' = \sum_{j=1}^{k} m_j' = \frac{s}{2}\sum_{j=1}^{k}\sum_{l=1}^{N} m_l\, sgn(x_j - x_l) = \frac{s}{2}\sum_{j=1}^{k} m_j \left(\sum_{l=1}^{j-1} m_l - \sum_{l=j+1}^{N} m_l\right).$$

On the other hand, $\sum_{l=j+1}^{N} m_l + \sum_{l=1}^{j-1} m_l + m_j = 0$, $j \geq 2$, i.e.

$$M_k' = \frac{s}{2}\sum_{j=1}^{k} m_j \left(m_j + 2\sum_{l=1}^{j-1} m_l\right) = \frac{s}{2}M_k^2, \tag{7.13}$$

as $M_k^2 = \sum_{j=1}^{k} m_j^2 + 2\sum_{l<j\leq k} m_j m_l$.

Conclusion 7.1. $M_k(t) = \frac{1}{c_k - \frac{s}{2}t}$ is C^∞ smooth for $t > \frac{2c_k}{s}$, where $M_k(0) = \frac{1}{c_k}$. We shall take $c_k < 0$ for each $k \in [1, N-1]$, c_k being different from each other. Put $const = \frac{2}{s}max_{1\leq k\leq N-1}c_k < 0 \Rightarrow M_k(t) < 0$ for $t > const$, $M_k(t) \approx -\frac{2}{st}$, $t \to +\infty$.

Define now $\Delta_k(t) = x_{k+1}(t) - x_k(t)$, $k = 1, 2, \ldots, N-1$. Evidently, $x_1' = -1/2\sum_{l=2}^{N} m_l(t)|x_1(t) - x_l(t)|$.

Moreover, $\Delta_k'(t) = x_{k+1}' - x_k' = \frac{1}{2}\sum_{l=1}^{N} m_l(|x_k - x_l| - |x_{k+1} - x_l|) = \frac{1}{2}[\sum_{l=1}^{k}(-m_l)(x_{k+1} - x_k) + \sum_{l=k+1}^{N} m_l(x_{k+1} - x_k)] = 1/2|x_{k+1} - x_k| \times (-\sum_{l=1}^{k} m_l + \sum_{l=k+1}^{N} m_l) = -\Delta_k \sum_{l=1}^{k} m_l$.

Thus, $\Delta_k' = -\Delta_k M_k$, as $\sum_{l=k+1}^{N} m_l = -\sum_{l=1}^{k} m_l$, $k = 1, 2, \ldots, N-1$.

Conclusion 7.2. $\Delta_k' = \frac{1}{\frac{s}{2}t - c_k}\Delta_k \Rightarrow \Delta_k(t) = A_k(\frac{st}{2} - c_k)^{2/s}$, $A_k = const > 0$, $t > \frac{2c_k}{s}$, $\Delta_k(0) = A_k(-c_k)^{2/s} > 0$ as $c_k < 0$.

Let us find now $m_k(t)$, $t > const$. As we know, $M_1 = m_1 = \frac{1}{c_1 - \frac{st}{2}} < 0$, $M_k - M_{k-1} = m_k$, $k \geq 2$ and $m_N = M_N - M_{N-1} = -M_{N-1} = \frac{1}{\frac{st}{2} - c_{N-1}} > 0$. Thus, $N - 1 \geq k \geq 2 \Rightarrow m_k = \frac{1}{c_k - \frac{st}{2}} - \frac{1}{c_{k-1} - \frac{st}{2}} = \frac{c_{k-1} - c_k}{(\frac{st}{2} - c_{k-1})(\frac{st}{2} - c_k)} \approx \frac{1}{s^2 t^2}$, $t \to \infty$.

As we know

$$\Delta_1 = x_2 - x_1 = A_1\left(\frac{st}{2} - c_1\right)^{2/s} \tag{7.14}$$

$$\Delta_2 = x_3 - x_2 = A_2\left(\frac{st}{2} - c_2\right)^{2/s}$$
$$\ldots$$
$$\Delta_{N-3} = x_{N-2} - x_{N-3} = A_{N-3}\left(\frac{st}{2} - c_{N-3}\right)^{2/s}$$
$$\Delta_{N-2} = x_{N-1} - x_{N-2} = A_{N-2}\left(\frac{st}{2} - c_{N-2}\right)^{2/s}$$
$$\Delta_{N-1} = x_N - x_{N-1} = A_{N-1}\left(\frac{st}{2} - c_{N-1}\right)^{2/s}.$$

By summing (7.14) we conclude that for each $j \geq 2$, $j \leq N$ we have:

$$x_j(t) = x_1 + \sum_{l=1}^{j-1} A_l \left(\frac{st}{2} - c_l \right)^{2/s}, \tag{7.15}$$

$t > \frac{2}{s} max_{1 \leq l \leq N-1} c_l = const.$

To complete our considerations we must find x_1 from the ODE

$$x_1' = -1/2 \sum_{l=2}^{N} m_l (x_l - x_1) \Rightarrow -2x_1' = \sum_{l=2}^{N} m_l \sum_{j=1}^{l-1} A_j \left(\frac{st}{2} - c_j \right)^{2/s} \tag{7.16}$$

As we mentioned, only two cases in (7.5) will be studied, namely I: $s = 1$ and II: $s = 2$.

2. In case I from the terms $m_l(x_l - x_1)$, $2 \leq l \leq N-1$ the following terms appear: $\frac{c_l - c_{l-1}}{(\frac{t}{2} - c_{l-1})(\frac{t}{2} - c_l)} \sum_{j=1}^{l-1} A_j (\frac{t}{2} - c_j)^2$, i.e. after the integration of (7.16) with respect to t we shall have a linear combination of t and $ln(\frac{t}{2} - c_j)$, i.e. $m_l(x_l - x_1)$, $2 \leq l \leq N - 1$ leads to a linear growth with respect to t in $x_1(t)$. As it concerns $m_N(x_N - x_1) = \frac{1}{\frac{t}{2} - c_{n-1}} \sum_{l=1}^{N-1} A_l (\frac{t}{2} - c_l)^2$ one can easily see that after integration with respect to t it gives rise to a quadratic term $\sum_{l=1}^{N-1} A_l (\frac{t}{2} - c_{N-1})^2$ plus linear term in t plus logarithmic term in $(\frac{t}{2} - c_{N-1})$.

Therefore,

$$x_1(t) = x_1(0) - 1/2 \sum_{l=1}^{N-1} A_l \left(\frac{t}{2} - c_{N-1} \right)^2 + B_1 t \tag{7.17}$$

+ logarithmic terms of the arguments $(\frac{t}{2} - c_j)$; $B_1 = const.$

Consequently,

$$x_1(t) \approx -1/2 \left(\sum_{l=1}^{N-1} A_l \right) \frac{t^2}{4} \tag{7.18}$$

for $t \to \infty$ and $x_j(t)$ has a quadratic growth at $+\infty$ for each j.

Then

$$l(t) = \frac{1}{2} \sum_{k=1}^{N} m_k(t) x_k(t) = \frac{1}{2} x_1(t) \frac{1}{c_1 - \frac{t}{2}} + \frac{1}{2} \frac{1}{\frac{t}{2} - c_{N-1}} \tag{7.19}$$

$$\times \left(x_1 + \sum_{1}^{N-1} A_j \left(\frac{t}{2} - c_j \right)^2 \right) + \frac{1}{2} \sum_{2}^{N-1} m_k(t) x_k(t)$$

$$= \frac{1}{2} x_1(t) \frac{c_1 - c_{N-1}}{(c_1 - \frac{t}{2})(\frac{t}{2} - c_{N-1})} + \frac{1}{2} \sum_{1}^{N-1} A_j \frac{(\frac{t}{2} - c_j)^2}{(\frac{t}{2} - c_{N-1})}$$

$$+\frac{1}{2}\sum_{k=2}^{N-1}m_k\left(x_1+\sum_{j=1}^{k-1}A_j\left(\frac{t}{2}-c_j\right)^2\right)\approx+1/2\sum_{j=1}^{N-1}A_j\frac{t}{2}>0.$$

for $t\to\infty$.

Certainly, for $j\geq 2$ and $t\to\infty$

$$x_j(t)\approx -1/2\left(\sum_{l=1}^{N-1}A_l\right)\frac{t^2}{4}+\sum_{l=1}^{j-1}A_l\frac{t^2}{4} \qquad (7.20)$$

$$=\frac{t^2}{4}\left(-\sum_{l=1}^{N-1}\frac{A_l}{2}+\sum_{l=1}^{j-1}A_l\right)=\frac{t^2}{8}\left(\sum_{l=1}^{j-1}A_l-\sum_{l=j}^{N-1}A_l\right).$$

More specially, if $j=N\geq 2$ then

$$x_N(t)\approx\frac{t^2}{8}\sum_{l=1}^{N-1}A_l, t\to\infty.$$

It is reasonable to impose the following conditions on the coefficients A_l in both cases $s=1$, $s=2$: for each $2\leq j\leq N$:

$$\sum_{l=1}^{j-1}A_l\neq\sum_{l=j}^{N-1}A_l \qquad (7.21)$$

and $A_l>0$.

Then the smooth curves $x_1(t)<x_2(t)<\ldots<x_N(t)$, $t>const$ are split into two parts: $x_1(t),\ldots,x_r(t)$ are such that $x_p(t)\to -\infty$ for $t\to +\infty$ and $x_{r+1}(t)<\ldots<x_N(t)$ are such that $x_p(t)\to +\infty$ for $t\to\infty$. The curve $x_1(t)$ belongs to the first class, while $x_N(t)$ belongs to the second one.

3. The case $s=2$ is similar to $s=1$ because the functions $m_k(t)$ are satisfying $M_k'=M_k^2$, i.e. $M_k=\frac{1}{c_k-t}$, $M_k(0)=\frac{1}{c_k}<0$, $t>c_k\Rightarrow M_k<0$, $m_1=\frac{1}{c_1-t}$, $m_j=\frac{c_{j-1}-c_j}{(t-c_j)(t-c_{j-1})}$, $2\leq j\leq N-1$ and $m_N=\frac{1}{t-c_{N-1}}$. Here $t>max_{1\leq j\leq N-1}c_j=const<0$. The equation $\Delta_k'=-\Delta_k M_k$ has the solution $\Delta_k(t)=A_k(t-c_k)$, $\Delta_k(0)=-A_k c_k$, $A_k>0$.

As in case I we can show (see (7.15)) that for $j\geq 2$, $j\leq N$

$$x_j=x_{j-1}+A_{j-1}(t-c_{j-1})=x_1+\sum_{l=1}^{j-1}A_l(t-c_l). \qquad (7.22)$$

Certainly,

$$x_1'=-1/2\sum_{l=2}^{N}m_l(x_l-x_1), t\geq const,$$

i.e.

$$x'_1 = -1/2 \sum_{l=2}^{N-1} m_l \sum_{j=1}^{l-1} A_j(t - c_j) - 1/2m_N \sum_{j=1}^{N-1} A_j(t - c_j). \qquad (7.23)$$

The integral of the first sum in the right-hand side of (7.23) gives rise of a linear combination of logarithmic terms: $ln(t-c_l)$, $ln(t-c_{l-1})$. The integral of the second sum in the right-hand side of (23) is: $-1/2 \sum_{l=1}^{N-1} A_l t - 1/2 \sum_{l=1}^{N-2} A_l(c_{N-1} - c_l)ln(t - c_{N-1})$.

Thus,

$$x_1(t) = x_1(0) - 1/2 \sum_{l=1}^{N-1} A_l t + \text{ linear combination of } ln(t - c_j), \qquad (7.24)$$

$1 \le j \le N - 1$.

According to (7.22) for $j \ge 2$

$$x_j(t) = x_1(t) + \sum_{l=1}^{j-1} A_l(t - c_l) \approx -1/2 \sum_{l=1}^{N-1} A_l t \qquad (7.25)$$

$$+ \sum_{l=1}^{j-1} A_l t = \frac{t}{2} \left(\sum_{l=1}^{j-1} A_l - \sum_{l=j}^{N-1} A_l \right), t \to +\infty.$$

In other words, for each $1 \le j \le N$ the function $x_j(t)$ has a linear growth (see condition (7.21)) at $+\infty$. In the case $j = N \Rightarrow x_N \approx \frac{1}{2} \sum_{l=1}^{N-1} A_l t$, $t \to \infty$. We remind of the reader that $\Gamma_j = \{(x,t) : x = x_j(t), t \ge const\}$, $1 \le j \le N$. According to (7.10) Γ_N is located above $\Gamma_{N-1} \ldots, \Gamma_2$ is located above Γ_1 in $\mathbf{R}_x \times \{t > const\}$ and Γ_j has asymptote for $t \to \infty$. The asymptotes q_j (straight lines) are monotonically increasing for $j = r + 1, \ldots, N$ and monotonically decreasing for $j = 1, \ldots, r$.

Our next step is to investigate $l(t)$ for $t \to \infty$. So $l(t) = \frac{1}{2}m_1 x_1 + \frac{1}{2}m_N x_N + \frac{1}{2}\sum_{l=2}^{N-1} m_l x_l = \frac{1}{2}x_1(m_1 + m_N) + \frac{1}{2}m_N \sum_{l=1}^{N-1} A_l(t - c_l) + o(1)$, $t \to \infty$, as $m_l \approx \frac{c_{l-1}-c_l}{t^2}$ for $t \to \infty$ and $2 \le l \le N - 1$.

Thus, $l(t) = \frac{1}{2}x_1 \frac{c_1-c_{N-1}}{(c_1-t)(t-c_{N-1})} + \frac{1}{2(t-c_{N-1})} \sum_{l=1}^{N-1} A_l(t - c_l) + o(1)$.

Consequently,

$$l(t) = \frac{1}{2} \sum_{l=1}^{N-1} A_l + o(1), t \to \infty, l(t) > 0. \qquad (7.26)$$

Geometrically, we have the following Fig. 7.1

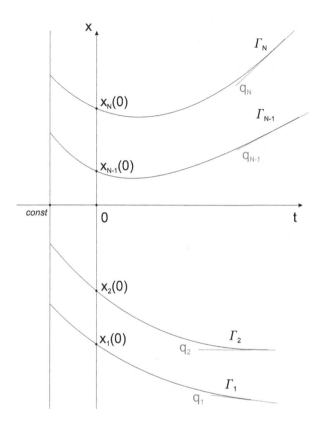

Fig. 7.1

7.2 Explicit formulas for the peakon-type solutions of the Hunter-Saxton equation

Our last step is to give an explicit expression for u in the following subdomains of $\mathbf{R}_x^1 \times \{t > const\}$: (i) above Γ_N, below Γ_1 and (ii) between Γ_j and Γ_{j+1}, $1 \le j \le N - 1$.

To do this we fix the point $D(x, t) \in \mathbf{R}_x^1 \times \{t > const\}$.

Case (i) Assume that D is located above Γ_N, i.e. $x > x_j(t)$ for each $1 \le j \le N$ (below Γ_1, i.e. $x < x_j(t), \forall j$). Then $u(D) = -1/2 \sum_{k=1}^{N} m_k(x - x_k) = \frac{1}{2} l(t)$ ($u(D) = -\frac{1}{2} l(t)$), $l(t) \to \frac{1}{2} \sum_1^{N-1} A_l > 0$, $t \to \infty$.

Case (ii). Suppose that D is located in the strip between Γ_j and Γ_{j+1}, i.e. $x_j(t) < x < x_{j+1}(t)$. Then $u(D) = -1/2 \sum_{l=1}^{j} m_l(x - x_l) +$

$\frac{1}{2}\sum_{l=j+1}^{N} m_l(x_l - x) = -1/2x(\sum_{l=1}^{j} m_l - \sum_{l=j+1}^{N} m_l) + \frac{1}{2}(-\sum_{l=j+1}^{N} m_l x_l + \sum_{l=1}^{j} m_l x_l) = -\frac{x}{2}P_j(t) + Q_j(t)$.

Having in mind that $\sum_{l=1}^{N} m_l(t) = 0 \Rightarrow \sum_{l=1}^{j} m_l = -\sum_{l=j+1}^{N} m_l$ we obtain $-\frac{x}{2}P_j(t) = -xM_j(t)$, $1 \le j \le N-1$.

Moreover,

$$Q_j = \frac{1}{2}\left(-\sum_{l=j+1}^{N-1} m_l x_l - m_N x_N + \sum_{l=2}^{j} m_l x_l + m_1 x_1\right)$$

$$= \frac{1}{2}(m_1 x_1 - m_N x_N) + o(1), t \to \infty.$$

Thus,

$$Q_j = \frac{1}{2}x_1(m_1 - m_N) - \frac{1}{2}m_N \sum_{l=1}^{N-1} A_l(t - c_l) + o(1) \qquad (7.27)$$

$$= \frac{1}{2}x_1 \frac{2t - c_1 - c_{N-1}}{(c_1 - t)(t - c_{N-1})} - \frac{1}{2}\sum_{l=1}^{N-1} A_l + o(1)$$

$$= \frac{1}{2}\sum_{l=1}^{N-1} A_l - 1/2 \sum_{l=1}^{N-1} A_l + o(1), t \to \infty.$$

This way we conclude that

$$u(D) = u(x,t) = \frac{x}{t - c_j} + Q_j(t), Q_j(t) \to_{t \to \infty} 0 \qquad (7.28)$$

if D is located between Γ_j and Γ_{j+1}, $1 \le j \le N-1$ (see Fig. 7.2).

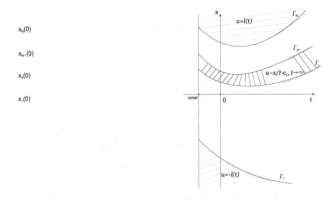

Fig. 7.2

Remark 7.1. The integral surface S of (7.5) is partially ruled (see [94] and Fig. 7.3), i.e. in each strip between Γ_j and Γ_{j+1} it has the form $u = A(t)x + B(t)$. Geometrically we have the following interpretation. Consider the smooth plane curve $\gamma : (t, 0, B(t)) \in 0_{tu}$ and the plane vector $\vec{l}\ (0, 1, A(t)) \in 0_{xu}$. Then in vector form S can be written as

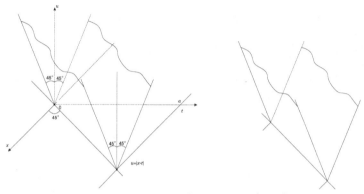

Fig. 7.3 $\quad u = |x - t|$.

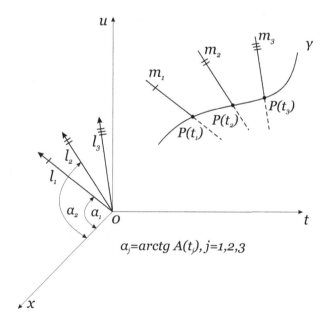

$$\alpha_j = arctg\, A(t_j), j=1,2,3$$

Fig. 7.4

$\overrightarrow{r} = P(t) + x \overrightarrow{l}\ (t)$, $x \in \mathbf{R}^1$, $t > const$; $P(t) = (t, 0, B(t))$. In other words through each point of γ is passing a straight line $m(t)$ collinear to the vector $\overrightarrow{l}$. The Causs curvature of S, $K = \frac{-(A')^2}{[1+A^2+(xA'+B')^2]^2} \leq 0$ (see Fig. 7.4).

It is interesting to point out that $A(t_0) = 0$ implies that $m_{t_0} || \overrightarrow{0_x}$, while the Gaussian curvature $K(t, x)$ of S vanishes at the critical points of the angle $\alpha(t) = arctg\ A(t)$ and for x - arbitrary, i.e. at the points at which $\alpha'(t) = A'(t) = 0$.

Chapter 8

δ-shocks for quasilinear hyperbolic systems in $\mathbf{R}^2$

8.1 Existence of δ-shocks

1. In this Chapter we shall deal at first with the Riemann problem for a special system of conservation laws. As it is well known from [69] (also see [20], [88], [70]), if the system is strictly hyperbolic, then the Riemann problem possesses a piecewise smooth weak solution. It can be constructed by using the method of characteristics or its slight modifications. Then shock waves and rarefaction waves appeared. So consider the system in $\{t > 0, x \in \mathbf{R}^1\}$

$$
\left|
\begin{array}{l}
\frac{\partial u}{\partial t} + \frac{\partial}{\partial x}(f(u)) = 0, \quad f \in C^2(\mathbf{R}^1), \\
\frac{\partial v}{\partial t} + \frac{\partial}{\partial x}(g(u)v) = 0, \quad g(u) = \frac{f(u)}{u}, \quad \text{for} \quad u \neq 0,
\end{array}
\right.
\tag{8.1}
$$

where $u(0, x) = \begin{cases} \tilde{d} \text{ for } x < 0 \\ c \text{ for } x > 0, \end{cases}$ $\tilde{d} > c > 0$, c, $\tilde{d}$ being constants and $v(0, x) = 1$.

It is interesting to point out that (8.1) generates the so called δ-shock wave for the Riemann problem and for $c \to 0$. We observe that the Cauchy problem

$$
\frac{\partial u}{\partial t} + (f(u))_x = 0, u|_{t=0} = u_0 \in C^1(\mathbf{R}^1),
$$

u_0, $u_0' \in L^\infty$ and for special nonlinearities $f(u)$ has a unique smooth solution u in a strip $[0, T) \ni t$ but "in general" u_x blows up for finite time $T > 0$. For example, if $f(u) = u^2$ then a global in time classical solution exists iff $u_0'(x) \geq 0$. Otherwise, one can find a formula for the life span time T of the solution u.

Proposition 8.1. [116].   *Consider the system (8.1) equipped with the above mentioned Riemann conditions. Assume that $f''(u) > 0$, $f(0) = 0$, $g(0) = f'(0)$. Then in the sense of Schwartz distributions there exists $\lim_{c \to 0} v(x,t) = 1 + t(g(\tilde{d}) - g(0))\delta_\Lambda$, δ being the delta function concentrated on the half line $\Lambda = \{(t,x) : x = g(\tilde{d})t, t > 0\}$.*

Remark 8.1. If $W = \begin{pmatrix} u \\ v \end{pmatrix}$ then (8.1) can be rewritten as:

$$\frac{\partial W}{\partial t} + \begin{pmatrix} f'(u) & 0 \\ g'(u)v & g(u) \end{pmatrix} \frac{\partial W}{\partial x} = 0, W|_{t=0} = \begin{pmatrix} \begin{pmatrix} \tilde{d}, & x < 0 \\ c, & x > 0 \end{pmatrix} \\ 1 \end{pmatrix}.$$

In fact, $g(u) = \frac{f(u)}{u} \in C^1(\mathbf{R})$.

The characteristic roots of our system are $\lambda = f'(u) = f'(0) + u \int_0^1 f''(su)ds$, $\mu = g(u) = \frac{f(u)}{u}$. Applying the Taylor's formula $f(u) = f(0) + uf'(0) + u^2 \int_0^1 (1-s)f''(su)ds$ we get that $\mu = \lambda + u \int_0^1 (-s)f''(su)ds$, i.e. $\mu \neq \lambda$ for $u \neq 0$.

2. Therefore, the above given system is strictly hyperbolic for $u \neq 0$ and possesses a unique weak solution as $\tilde{d} > c > 0$. In order to construct at first u and then v as piecewise constant functions we shall propose a small excursion in the conservation laws theory: [69], [71], [65], [39], [72], [47]. Thus, consider Fig. 8.1, where $\Gamma : x = x(t), t \geq 0$ is a smooth curve,

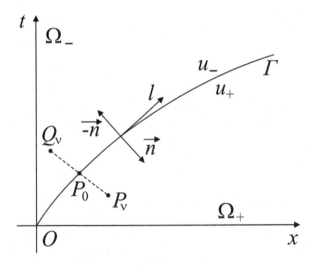

Fig. 8.1

$\Omega = \{(t,x),\ t > 0\}$, $\Omega = \Omega_+ \cup \Omega_- \cup \Gamma$ and $\Omega_\pm$ are open domains. We assume that $u \in C^1(\Omega_-) \cap C^1(\Omega_+)$, $u \in C^0(\bar{\Omega}_+ \setminus \{x > 0\}) \cap C^0(\bar{\Omega}_- \setminus \{x < 0\})$ and we denote $u_+(P_0) = \lim_{\substack{P_\nu \to P_0 \in \Gamma \\ P_\nu \in \Omega_+}} u(P_\nu)$, $u_-(P_0) = \lim_{\substack{Q_\nu \to P_0 \in \Gamma \\ Q_\nu \in \Omega_-}} u(Q_\nu)$, $[u](P_0) = u_+(P_0) - u_-(P_0)$. Evidently, $\vec{l} = (1, x'(t))$ is the tangential vector to Γ at the point $(t, x(t))$ and $\vec{n} = \left(\frac{-x'}{\sqrt{1+(x')^2}}, \frac{1}{\sqrt{1+(x')^2}}\right)$ is the unit normal to Γ at $(t, x(t))$ pointing inside Ω_+. According to the jump formula for the Schwartz distribution $u \in D'(t > 0)$, $u_t \in D'$ (see [117] and Chapter V):

$$u_t = \{u_t\} + [u]_\Gamma \cos(n,t)\delta_\Gamma \tag{8.2}$$

$$(f(u))_x = \{(f(u))_x\} + [f(u)]_\Gamma \cos(n,x)\delta_\Gamma,$$

where $[f(u)]_\Gamma = f(u_+) - f(u_-)$ and $\{u_t\}$ stands for the distribution generated by the measurable function u_t which is defined almost everywhere (outside Γ, evidently). If $u_t + (f(u))_x = 0$ in classical sense outside Γ, i.e. in $\Omega \setminus \Gamma$, then $u_t + (f(u))_x = 0$ in distribution sense in Ω (i.e. in weak sense) iff

$$[u]_\Gamma \cos(n,t) + [f(u)]_\Gamma \cos(n,x) = 0, \tag{8.3}$$

i.e. iff

$$x'(t) = \frac{[f(u)]_\Gamma}{[u]_\Gamma}, \forall t > 0. \tag{8.4}$$

(8.4) is the well known Rankine-Hugoniot condition, $\cos(n,t) = -\frac{x'}{\sqrt{1+(x')^2}}$, $\cos(n,x) = \frac{1}{\sqrt{1+(x')^2}}$.

Concerning the jump formula (8.2) we have the following additional comment. According to the corresponding formulas from Analysis for each test function $\varphi(t,x)$

$$< \delta_\Gamma, \varphi > = \int_\Gamma \varphi \, dl = \int_0^\infty \varphi(t, x(t))\sqrt{1+(x')^2} \, dt,$$

as in the curvilinear integral along Γ we have $dl = \sqrt{1+(x')^2} \, dt$. Therefore,

$$< [u]_\Gamma \cos(n,t)\delta_\Gamma, \varphi > = -\int_0^\infty [u]_\Gamma x'(t)\varphi(t, x(t)) \, dt$$

and

$$< [u]_\Gamma \cos(n,x)\delta_\Gamma, \varphi > = \int_0^\infty [u]_\Gamma \varphi(t, x(t)) \, dt.$$

Summarizing (8.1), (8.2), (8.3), (8.4) we come to the following definition of the generalized (weak) solution of the Cauchy problem

$$u_t + (f(u))_x = 0 \qquad (8.5)$$

$$u|_{t=0} = u_0(x) \in L^1_{loc}(\mathbf{R}^1).$$

Definition 8.1. We shall say that $u \in L^1_{loc}([0,\infty) \times \mathbf{R}^1_x)$ with $f(u) \in L^1_{loc}([0,\infty) \times \mathbf{R}^1_x)$ is a weak solution of (8.5) iff the integral identity

$$\int_0^\infty \int_{-\infty}^\infty (u\varphi_t + f(u)\varphi_x)dt\,dx + \int_{-\infty}^\infty u_0(x)\varphi(0,x)dx = 0$$

holds for each test function $\varphi \in C_0^1([0,\infty) \times \mathbf{R}^1_x)$.

Equivalently, $\varphi \in D([0,\infty) \times \mathbf{R})$, D containing the C^∞ test functions. Certainly, $\varphi(0,x) \in C_0^\infty(\mathbf{R})$ and $\varphi(0,x)$ is not obliged to be identically 0.

In the special case $v_t + (G(u,v))_x = 0$, $G(u,v) = g(u)v$, the Rankine-Hugoniot condition takes the form $x'(t) = \frac{[G(u,v)]_\Gamma}{[v]_\Gamma}$, where $[G(u,v)]_\Gamma = G(u_+,v_+) - G(u_-,v_-)$. It is assumed that v has a finite jump along Γ.

Going back to the first equation of (8.1) equipped with the corresponding Cauchy data we obtain that

$$\left| \begin{array}{ll} x'(t) = \frac{f(u_+)-f(u_-)}{u_+-u_-}, & u_+ = c \\[2mm] x(0) = 0, & u_- = \tilde{d}. \end{array} \right.$$

Thus,

$$u_c = \begin{cases} \tilde{d}, & x < x(t) \\ c, & x > x(t), \end{cases} \qquad (8.6)$$

where $\Gamma : x(t) = \frac{f(\tilde{d})-f(c)}{\tilde{d}-c}t$.

In fact, the constants c, $\tilde{d}$ satisfy in classical sense our equation in Ω but outside Γ and the Rankine-Hugoniot condition holds at Γ.

There are no difficulties in solving the second equation of (8.1) with unknown function v. As we know the number of discontinuity curves is 2. We have found Γ. Now we are looking for the second one $\Lambda = \{x = y(t), t > 0\}$. The solution v should have discontinuities along $\Lambda \subset \{t > 0\}$ and Γ and it is equal to (different) constants in Ω but outside Λ and Γ.

These are two trivial observations: $0 < c < \tilde{d}$

$$\Rightarrow \frac{f(c)}{c} < \frac{f(\tilde{d})}{\tilde{d}} < \frac{f(\tilde{d})-f(c)}{\tilde{d}-c}. \qquad (8.7)$$

They are standard properties of the strictly convex functions. In fact, $\left(\frac{f(u)}{u}\right)' = \frac{uf'(u)-f(u)}{u^2}$ for $u > 0$. But we have proved above that $\frac{f(u)}{u} - f'(u) = -u\int_0^1 sf''(su)ds < 0$ for $u > 0$ as $f'' > 0$. Therefore, $\frac{f(u)}{u}$ is strictly monotonically increasing function for $u > 0$. The inequality $0 < \frac{f(\tilde{d})}{\tilde{d}} < \frac{f(\tilde{d})-f(c)}{\tilde{d}-c}$ is equivalent to $\frac{f(\tilde{d})}{\tilde{d}} > \frac{f(c)}{c} > 0$ etc.

Denote $\Omega_+ = \{(t,x) : x > x(t) \equiv \frac{f(\tilde{d})-f(c)}{\tilde{d}-c}t, t > 0\}$, $\Omega_- = \{(t,x) : x < x(t), t > 0\}$. Suppose then that $\Lambda \subset \Omega_+$. Then the Rankine-Hugoniot condition along Λ and for the second equation of the system (8.1) implies that $\frac{dy}{dt} = \frac{[g(u)v]_\Lambda}{[v]_\Lambda}$, $y(0) = 0$, i.e. $y' = \frac{[g(c)v]_\Lambda}{[v]_\Lambda} = g(c)$ and consequently according to (8.7) $y(t) = g(c)t = \frac{f(c)}{c}t < \frac{f(\tilde{d})}{\tilde{d}}t < x(t)$ for $t > 0 \Rightarrow$ $y(t) < x(t) \Rightarrow \Lambda \subset \Omega \setminus \Omega_+$. Therefore, the discontinuity curve $\Lambda \subset \Omega_-$ - contradiction. Hence, $\frac{dy}{dt} = \frac{[g(u)v]_\Lambda}{[v]_\Lambda} = g(\tilde{d}) \Rightarrow y(t) = g(\tilde{d})t, \forall t > 0$. Put $v(t,x) = \alpha$ in the angle $\{(t,x) : y(t) < x < x(t), t > 0\}$ and $v \equiv 1$ outside this angle. The Rankine-Hugoniot condition holds along Λ, while the same condition along Γ takes the form:

$$\frac{dx}{dt} = \frac{f(\tilde{d})-f(c)}{\tilde{d}-c} = \frac{[g(u)v]_\Lambda}{[v]_\Lambda} = \frac{g(c)-g(\tilde{d})\alpha}{1-\alpha} \Rightarrow \qquad (8.8)$$

$\alpha = \frac{\tilde{d}}{c}$, as $g(c) = \frac{f(c)}{c}$, $g(\tilde{d}) = \frac{f(\tilde{d})}{\tilde{d}}$ (see Fig. 8.2).

So $v_c = \begin{cases} 1, & x < y(t) \\ \tilde{d}/c, & y(t) < x < x(t) \\ 1, & x > x(t) \end{cases}$ satisfies the second equation of (8.1) in weak sense as the constants satisfy the equation outside Λ and Γ and the Rankine-Hugoniot conditions are satisfied along Λ and Γ.

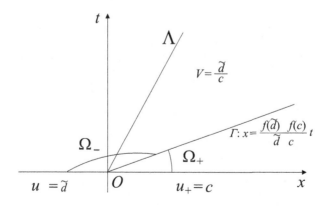

Fig. 8.2

3. We are ready now to prove Proposition 8.1.

Let $\varphi \in C_0^\infty(\Omega)$. Then

$$< v_c, \varphi > = \int_0^\infty \left(\int_{-\infty}^{y(t)} \varphi(t,x)dx + \int_{y(t)}^{x(t)} \frac{\tilde{d}}{c}\varphi(t,x)dx + \int_{x(t)}^\infty \varphi(t,x)dx \right) dt. \tag{8.9}$$

It is obvious that there exists a constant $0 < K(\varphi) < \infty$ and such that in (8.9) we are integrating with respect to t in the finite interval $[0, K]$ (not in $[0, +\infty)$). Thus, $lim_{c\to+0}x(t) = \frac{f(\tilde{d})}{\tilde{d}}t = y(t)$ for each $t \geq 0$.

Geometrically the family of straight lines $\Gamma_c : (t, x(t))$ tends for $c \to 0$ to the straight line $\Lambda : (t, y(t))$. Then

$$\int_{-\infty}^{y(t)} \varphi(t,x)dx + \int_{x(t)}^{+\infty} \varphi(t,x)dx \to \int_{-\infty}^\infty \varphi(t,x)dx \tag{8.10}$$

for $c \to +0$.

We shall use the identity

$$\frac{\tilde{d}}{c} \int_{y(t)}^{x(t)} \varphi(t,x)dx = \frac{\tilde{d}}{c} \int_{y(t)}^{x(t)} (\varphi(t,x) - \varphi(t, g(\tilde{d})t))dx \tag{8.11}$$

$$+\frac{\tilde{d}}{c}(x(t) - y(t))\varphi(t, g(\tilde{d})t) \equiv I_c + II_c.$$

On the other hand,

$$lim_{c\to0}\frac{\tilde{d}}{c}(x(t) - y(t)) = \tilde{d}\, t\, lim_{c\to0}\frac{1}{c}\left(\frac{f(\tilde{d}) - f(c)}{\tilde{d} - c} - \frac{f(\tilde{d})}{\tilde{d}} \right) \tag{8.12}$$

$$= t \lim_{c\to0}\left(\frac{f(\tilde{d})}{\tilde{d} - c} - \frac{\tilde{d}}{\tilde{d} - c}\cdot\frac{f(c)}{c} \right) = t(g(\tilde{d}) - g(0)),$$

as $g(0) = f'(0) = lim_{c\to0}\frac{f(c)}{c}$.

According to the mean-value theorem

$$|\varphi(t,x) - \varphi(t, g(\tilde{d})t| \leq |x - g(\tilde{d})t| \tag{8.13}$$

$$\times max_{0\leq t, y(t)\leq x\leq x(t)}\left| \frac{\partial\varphi}{\partial x}(t, g(\tilde{d})t + \Theta(x - g(\tilde{d})t)) \right|, 0 < \Theta < 1,$$

and $0 \leq x - g(\tilde{d})t \leq x(t) - y(t)$.

Combining (8.11), (8.12) and (8.13) we get that $lim_{c\to0}I_c = 0$ for each $t > 0$ and $lim_{c\to0}II_c = t(g(\tilde{d}) - g(0))\varphi(t, g(\tilde{d})t)$.

According to (8.9)

$$lim_{c\to 0} < v_c, \varphi >= \int_0^\infty \int_{-\infty}^\infty \varphi(t,x)dt\,dx + \int_0^\infty t(g(\tilde{d}) - g(0))\varphi(t, g(\tilde{d})t)dt.$$

Therefore, in the weak topology of $D'(\Omega)$

$$v = lim_{c\to 0}v_c = 1 + t(g(\tilde{d}) - g(0))\delta_\Lambda, \tag{8.14}$$

where $< \delta_\Lambda, \varphi >= \int_0^\infty \varphi(t, g(\tilde{d})t)dt$; $\Lambda = \{(t,x) : t > 0, x = g(\tilde{d})t\}$.

Certainly,

$$lim_{c\to 0}u_c = \begin{cases} \tilde{d}, & x < y(t) \\ 0, & x > y(t). \end{cases} \tag{8.15}$$

We shall denote $lim_{c\to 0}u$ again by u.

The following interesting question arises: Do $(u, v) \in D'(\Omega)$ and defined by (8.14), (8.15) satisfy the system (8.1) and in which sense? This is a nontrivial question as the derivative $\frac{\partial}{\partial x}(uv)$ participates in (8.1), the function u has a discontinuity along Λ and $v - 1 = t\delta_\Lambda$. As we know, multiplication of such distributions is impossible in the frames of Schwarts distribution theory. In fact, if $H(x)$ is the one dimensional Heaviside function and $\delta(x)$ is the standard Dirac delta function then the existence of $H(x)\delta(x) \in D'(\mathbf{R}^1)$ and the rule for differentiation of product would imply $(H\delta)' = \delta^2 + H\delta''$. Certainly, $\delta^2(x)$ does not exist in $D'(\mathbf{R}^1)$. There are two ways to avoid this difficulty, namely, to use the approach of Colombeau [28] (see also [86]) or to propose a suitable definition of the generalized solution of quasilinear systems via integral identities as it was done in Definition 8.1. This approach was developed by Danilov-Shelkovich (see [41], [42]) and we shall follow it with unsignificant simplifications in order to prove that (8.14), (8.15) satisfy (8.1) in appropriate sense. Of course, the integral identity mentioned above must be "equivalent" in some appropriate sense to the quasilinear system (8.1) equipped with the corresponding initial data. In [106], [107] Shelkovich used the vanishing viscosity approach in his considerations (see also [60]), while we apply here the method of characteristics. Interesting results on the Cauchy problem for quasilinear systems are given in [66] and [122].

4. Suppose that γ is a smooth curve located in the upper half plane $\{t > 0\}$, $\gamma : x = \gamma(t)$ and let γ start from (say) the origin 0. We denote by $l(1, \gamma'(t))$ the tangential vector to γ at the point $(t, \gamma(t))$ and by $\frac{\partial}{\partial l}$ the corresponding directional derivative. Thus, if $\varphi(t, x) \in D([0, \infty) \times \mathbf{R}_x^1)$ is a test function, then $\frac{\partial \varphi}{\partial l}|_\gamma = \frac{\partial \varphi}{\partial t}(t, \gamma(t)) + \frac{\partial \varphi}{\partial x}(t, \gamma(t)).\gamma'(t) = \frac{d}{dt}\varphi(t, \gamma(t))$. Further on we shall assume that $|l_1| = 1 \Rightarrow l_1 = (\frac{1}{\sqrt{1+(\gamma')^2}}, \frac{\gamma'}{\sqrt{1+(\gamma')^2}})$.

We omit the obvious modification in the definition of $\frac{\partial \varphi}{\partial l_1}\big|_\gamma$ leading to: $\frac{\partial \varphi}{\partial l_1}\big|_\gamma = \frac{(\varphi(t,\gamma(t)))'}{\sqrt{1+(\gamma')^2}}$. Consider now the nonlinear system

$$\left|\begin{array}{l} u_t + (F(u,v))_x = 0,\ u|_{t=0} = u^0(x) \in L^\infty(\mathbf{R}) \\ v_t + (G(u,v))_x = 0,\ v|_{t=0} = \hat{v}^0(x) \in L^\infty(\mathbf{R}) \end{array}\right. \tag{8.16}$$

where F, G are smooth functions linear with respect to v.

Definition 8.2. (see [106]). A pair of Schwartz distributions $(u(t,x),$ $v(t,x))$ and a smooth curve γ crossing the $\vec{0_x}$ axes at some point A, where

$$v(t,x) = \hat{v}(t,x) + e(t,x)\delta_\gamma, \tag{8.17}$$

$$u,\hat{v} \in L^\infty([0,\infty) \times \mathbf{R}^1_x), e|_\gamma \in C^1,$$

are called a generalized δ-shock wave type solution of (8.16) if the integral identities

$$\int_0^\infty \int_{-\infty}^\infty (u\varphi_t + F(u,\hat{v})\varphi_x)dt\,dx + \int_{-\infty}^\infty u^0(x)\varphi(0,x)dx = 0, \tag{8.18}$$

$$\int_0^\infty \int_{-\infty}^\infty (\hat{v}\varphi_t + G(u,\hat{v})\varphi_x)dt\,dx + \int_\gamma e(t,x)\frac{\partial \varphi}{\partial l_1}dl \tag{8.19}$$

$$+ \int_{-\infty}^\infty \hat{v}^0(x)\varphi(0,x)dx + e^0\varphi(A) = 0, e^0 = const.$$

hold for all $\varphi \in D([0,\infty) \times \mathbf{R}^1_x)$.

Evidently, $dl = \sqrt{1+(\gamma')^2} \Rightarrow \int_\gamma e(t,\gamma(t))\frac{\partial \varphi}{\partial l_1}dl = \int_0^\infty e(t)\frac{d}{dt}\varphi(t,\gamma(t))dt$.

The definition of delta function δ_γ concentrated on the smooth curve γ is standard, $v = \hat{v}$ outside γ.

Let the upper half-plane $\{t > 0\}$ be cut by γ, $\gamma \ni 0$ into a left and right-hand parts $\Omega_\pm = \{(t,x) : \pm(x - \gamma(t)) > 0\}$. Similar things were done in proving Proposition 8.1 of this Chapter. Certainly, Ω_+ is the right hand part and let $(u(t,x),v(t,x)),\gamma$ be a generalized δ-shock wave type solution of (8.16) and such that $u,\hat{v} \in C^1(\Omega_\pm)$, $u,\hat{v} \in C^0(\bar{\Omega}_+) \cap C^0(\bar{\Omega}_-)$. Then it can be shown (see [106]) that the following Rankine-Hugoniot conditions hold for δ shocks:

$$\gamma'(t) = \frac{[F(u,\hat{v})]_\gamma}{[u]_\gamma}, \tag{8.20}$$

$$\frac{d}{dt}(e(t,\gamma(t))) = \frac{[F(u,\hat{v})]_\gamma}{[u]_\gamma}[\hat{v}]_\gamma - [G(u,\hat{v})]_\gamma. \tag{8.21}$$

The condition (8.20) is well-known, while the right-hand side of (8.21) is called the Rankine-Hugoniot deficit. Usually, we shall work with $F(u, v) = f(u)$, $G(u, v) = g(u)v$, $g(u) = \frac{f(u)}{u}$, $f(0) = 0$.

Then

$$\frac{d}{dt}(e(t, \gamma(t))) = \frac{[f(u)]_\gamma}{[u]_\gamma}[\hat{v}]_\gamma - [g(u)\hat{v}]_\gamma. \qquad (8.22)$$

As usual $[\hat{v}]_\gamma = \hat{v}_+ - \hat{v}_-$ on γ.

We shall prove the deficit formula (8.22) as it is a good exercise of the distribution theory. Consider at first the integral identity (19) and put $\varphi \in C_0^\infty(\Omega \setminus \gamma)$. Then

$$0 = \int_0^\infty \int_{-\infty}^\infty (\hat{v}\varphi_t + G(u, \hat{v})\varphi_x)dx\,dt = <\hat{v}, \varphi_t >$$

$$+ < g(u)\hat{v}, \varphi_x >= - < \hat{v}_t + (g(u)\hat{v})_x, \varphi >$$

(the latter identity being fulfilled in distribution sense). Therefore, $\hat{v}_t + (g(u)\hat{v})_x = 0$ in $\Omega \setminus \Gamma$ in a classical sense as $u, \hat{v} \in C^1(\Omega_\pm)$. Therefore, $\{\hat{v}_t + (g(u)\hat{v})_x\} = 0$.

Assume now that $\varphi \in C_0^\infty(\Omega)$. Then in a distribution sense and applying again (8.19) we have:

$$0 =< \hat{v}, \varphi_t > + < g(u)\hat{v}, \varphi_x > + \int_\gamma e\frac{\partial\varphi}{\partial l_1}dl$$

$$= - < \hat{v}_t, \varphi > - < (g(u)\hat{v})_x, \varphi > + \int_\gamma e\frac{\partial\varphi}{\partial l_1}dl.$$

The well known jump formulas (8.2) and (8.3) give us that $\gamma'(t) = \frac{[f(u)]_\gamma}{[u]_\gamma}$ and

$$\hat{v}_t + (g(u)\hat{v})_x = \{\hat{v}_t + (g(u)\hat{v})_x\} + [\hat{v}]_\Gamma cos(n, t)\delta_\gamma$$

$$+ [g(u)\hat{v}]_\gamma cos(n, x)\delta_\gamma \Rightarrow \int_0^\infty \varphi(t, \gamma(t))([g(u)\hat{v}]_\gamma$$

$$- \gamma'[\hat{v}]_\gamma)dt =< \hat{v}_t + (g(u)\hat{v})_x, \varphi > .$$

Consequently,

$$\int_0^\infty \varphi(t, \gamma(t))([g(u)\hat{v}]_\gamma - \gamma'[\hat{v}]_\gamma)dt$$

$$= \int_0^\infty e(t, x(t))\frac{d}{dt}\varphi(t, \gamma(t))dt = - \int_0^\infty \varphi(t, \gamma(t))\frac{d}{dt}e(t, x(t))dt,$$

as $\varphi(0, 0) = 0$. Having in mind that $\varphi \in C_0^\infty(\Omega)$ is arbitrary we get (8.22).

5. Now we are ready to prove that (8.14), (8.15) satisfy (8.1) in the sense of Definition 8.2. As the proof of (8.18) is standard we shall verify (8.19) only with $\hat{v} = 1$, $e(t) = (g(\tilde{d}) - g(0))t$, $e(0) = 0$, $v = 1$ outside $\Lambda \equiv \gamma$. We observe that (8.22) $\Rightarrow \frac{de}{dt} = -[g(u)]_\Lambda = g(\tilde{d}) - g(0)$ etc. The system (8.1) is not strictly hyperbolic as $u = 0$ for $x > y(t)$ (see Remarks 8.1 and 8.5). Thus, we shall show (8.19) with $G(u,v) = g(u)v$ and for each test function φ, i.e.

$$\int_0^\infty \int_{-\infty}^\infty (\varphi_t + g(u)\varphi_x)dx\, dt + \int_0^\infty (g(\tilde{d}) - g(0))t\frac{d}{dt}\varphi(t, \tilde{d}t)dt \qquad (8.23)$$

$$+ \int_{-\infty}^\infty \varphi(0,x)dx \equiv I + II + III = 0(A = 0, e^0 = e(A) = 0).$$

By integration by parts we get

$$II = -(g(\tilde{d}) - g(0))\int_0^\infty \varphi(t, \tilde{d}t)dt. \qquad (8.24)$$

As it concerns I, we shall use the following identity: $\frac{d}{dt}\int_{-\infty}^{\tilde{d}t} \varphi(t,x)dx = \int_{-\infty}^{\tilde{d}t} \varphi_t(t,x)dx + \tilde{d}\varphi(t, \tilde{d}t)$. Then

$$I = \int_0^\infty \left(\int_{-\infty}^{\tilde{d}t} (\varphi_t + g(\tilde{d})\varphi_x)dx \right) dt + \int_0^\infty \left(\int_{\tilde{d}t}^{+\infty} (\varphi_t + g(0)\varphi_x)dx \right) dt$$

$$= \int_0^\infty \frac{d}{dt} \left(\int_{-\infty}^{\tilde{d}t} \varphi(t,x)dx \right) dt - \tilde{d}\int_0^\infty \varphi(t, \tilde{d}t)dt$$

$$+ g(\tilde{d})\int_0^\infty \varphi(t, \tilde{d}t)dt + \int_0^\infty \frac{d}{dt} \left(\int_{\tilde{d}t}^{+\infty} \varphi dx \right) dt + \tilde{d}\int_0^\infty \varphi(t, \tilde{d}t)dt$$

$$-g(0)\int_0^\infty \varphi(t, \tilde{d}t)dt \Rightarrow$$

$$I = \int_{-\infty}^\infty \varphi(\infty, x)dx - \int_{-\infty}^0 \varphi(0,x)dx + (g(\tilde{d}) - g(0))\int_0^\infty \varphi(t, \tilde{d}t)dt \quad (8.25)$$

$$- \int_0^\infty \varphi(0,x) = -\int_{-\infty}^\infty \varphi(0,x)dx + (g(\tilde{d}) - g(0))\int_0^\infty \varphi(t, \tilde{d}t)dt.$$

Summing up (8.24), (8.25) and III we obtain

$$I + II + III = 0,$$

i.e. (8.23) holds.

Consequently, (u, v) and Λ form a generalized δ-shock wave type solution of (8.1).

The last thing is to motivate unformally Definition 8.2 on the level of the system (8.1). To do this we consider the identity $< v_t, \varphi > +(g(u)v)_x, \varphi >= 0,\ \varphi \in D$. Then it is equivalent to $0 =< v, \varphi_t > + < g(u)v, \varphi_x >$ up to the summand $\int_{-\infty}^{\infty} \varphi(0, x)v(0, x)dx$. Having in mind that $v = 1 + e(t)\delta_\Lambda$, where $e(t) = (g(t) - g(0))t$ we obtain

$$< 1, \varphi_t + g(u)\varphi_x > + < \delta_\Lambda, e(t)(\varphi_t + g(u)\varphi_x) >$$

$$= \int_0^\infty \int_{-\infty}^\infty (\varphi_t + g(u)\varphi_x)dx\,dt + < \delta_\Lambda, e(t)(\varphi_t + g(u)\varphi_x) > .$$

According to (8.15) u is not defined at Λ but there exist $u_+ = 0$ and $u_- = \tilde{d}$ along Λ. Therefore, we could define $g(u)|_\Lambda$ as $g(u_-)$ or as $g(u_+) = f'(u_+)$. It is reasonable to take $g(u)|_\Lambda = g(u_-) = g(\tilde{d}) = y'(t)$, where $\Lambda : y(t) = g(\tilde{d})t$. Then

$$(\varphi_t + g(u)\varphi_x)|_\Lambda = \varphi_t(x, y(t)) + y'(t)\varphi_x(x, y(t))$$

$$= \frac{d}{dt}\varphi(t, y(t)) = \frac{d}{dt}(\varphi|_\Lambda) = \frac{\partial\varphi}{\partial l}|_\Lambda \Rightarrow \int_0^\infty \int_{-\infty}^\infty (\varphi_t + g(u)\varphi_x)dx\,dt$$

$$+ \int_\Lambda e(t)\frac{\partial\varphi}{\partial l_1}dl = 0.$$

This way we come in a natural way to (8.19) with $G(u, v) = g(u)v$, $\hat{v} = 1$, $v = 1 + e(t)\delta_\Lambda$, $l = (1, g(\tilde{d}))$, $\Lambda : y(t) = g(\tilde{d})t$, $e(t) = (g(\tilde{d}) - g(0))t$.

6. We shall complete this section by investigating the case for Riemann problems with $\tilde{d} > 0$, $\tilde{d} > c$.

Definition 8.3. Consider the system (8.16) with Cauchy data $(u^0(x), v^0(x)) \in L_{loc}^\infty(\mathbf{R}^1)$. We shall say that $(u, v) \in L_{loc}^\infty(\mathbf{R}_+^2)$ is a weak solution of (8.16) if the following two integral identities hold for each test function $\varphi \in C_0^1(\mathbf{R}_+^2)$, $\mathbf{R}_+^2 = [0, \infty) \times \mathbf{R}_x^1$:

$$\int_0^\infty \int_{-\infty}^\infty (u\varphi_t + F(u, v)\varphi_x)dx\,dt + \int_{-\infty}^\infty u^0(x)\varphi(0, x)dx = 0 \qquad (8.26)$$

$$\int_0^\infty \int_{-\infty}^\infty (v\varphi_t + G(u, v)\varphi_x)dx\,dt + \int_{-\infty}^\infty v^0(x)\varphi(0, x)dx = 0. \qquad (8.27)$$

Suppose now that

$$F(u,v) = f(u), f(0) = 0, f'' > 0 \text{ everywhere and } G(u,v) = g(u)v, g \in C^1 \tag{8.28}$$

and consider the system (8.16), (8.28) equipped with constant initial data:

$$u^0(x) = \begin{cases} u_-, & x < 0 \\ u_+, & x > 0, \end{cases} u_+ > u_-, v^0(x) = \begin{cases} v_-, & x < 0 \\ v_+, & x > 0. \end{cases} \tag{8.29}$$

We are interested in the answer of the question: under which conditions the so called rarefaction wave solution of the equation $u_t + (f(u))_x = 0$, $u = u^0$ ((8.29)) and v form a weak solution of (8.16), (8.29). To do this we shall construct the rarefaction solution at first. Thus, according to the definition of rarefaction wave [63] let $u = \psi(\frac{x}{t})$ be a classical solution of $u_t + (f(u))_x = 0$, $f'' > 0$. Then $f'(\psi(p)) = p$. Having in mind that f' is strictly monotonically increasing as $(f')' > 0$ we conclude that f' is invertible and $\psi(p) = (f')^{-1}(p) \in C^1$. Define then

$$u(t,x) = \begin{cases} u_-, & x \leq f'(u_-)t \\ \psi(\frac{x}{t}), & f'(u_-)t < x < f'(u_+)t \\ u_+, & x \geq f'(u_+)t \end{cases} \tag{8.30}$$

and

$$v(t,x) = \begin{cases} v_-, & x \leq f'(u_-)t \\ 0, & f'(u_-)t < x < f'(u_+)t \\ v_+, & x \geq f'(u_+)t. \end{cases} \tag{8.31}$$

Evidently, $u \in C(\bar{\mathbf{R}}^2_+ \setminus \{0\})$ and jointly with the fulfilment of the Rankine-Hugoniot condition (8.3) ($[u]_\Gamma = 0$, $[f(u)]_\Gamma = 0$) we obtain that (8.30) is a weak solution (called rarefaction wave). Our next step is to write (8.27) into an appropriate form taking into account (8.28), (8.29), (8.30), (8.31). As we have done similar computations in (8.23)-(8.25) here a short sketch will be proposed only.

Thus,

$$\Delta = \int_0^\infty \int_{-\infty}^\infty v(\varphi_t + g(u)\varphi_x)dx\,dt + v_- \int_{-\infty}^0 \varphi(0,x)dx \tag{8.32}$$

$$+ v_+ \int_0^\infty \varphi(0,x)dx = \int_0^\infty \left(\int_{-\infty}^{f'(u_-)t} v_-(\varphi_t + g(u_-)\varphi_x)dx \right) dt$$

$$+ \int_0^\infty \left(\int_{f'(u_+)t}^\infty v_+(\varphi_t + g(u_+)\varphi_x)dx \right) dt + v_- \int_{-\infty}^0 \varphi(0,x)dx$$

$$+ v_+ \int_0^\infty \varphi(0,x)dx = v_-(f'(u_-) - g(u_-)) \int_0^\infty \varphi(t, f'(u_-)t)dt$$

$$+ v_+(f'(u_+) - g(u_+)) \int_0^\infty \varphi(t, f'(u_+)t)dt, \; f'(u_+) > f'(u_-)$$

as $u_+ > u_-$ and f' is strictly monotonically increasing function.

Proposition 8.2. *Consider the system (8.16) under the conditions (8.28), (8.29). Then it possesses a weak solution of the form (8.30), (8.31) iff* $v_-(f'(u_-) - g(u_-)) = 0$ *and* $v_+(f'(u_+) - g(u_+)) = 0$.

Certainly, the case $v_+ = v_- = 0$ is trivial as it implies $v \equiv 0$. If v_+, v_- are both different from 0, then our condition takes the form $f'(u_\pm) - g(u_\pm) = 0$.

Corollary 8.1. *Consider the system (8.16) under the conditions (8.28), (8.29) and assume that* $f'(u) = g(u)$ *everywhere. Then for arbitrary initial data* u^0, v^0, $u_+ > u_-$ *our system possesses a weak solution of the from (8.30), (8.31).*

Therefore, in this very special case a rarefaction wave appears but a generalized δ-shock wave is not generated.

Remark 8.2. De facto we have shown in (8.32) that v satisfies the "integral identity"

$$\int_0^\infty \int_{-\infty}^\infty v(\varphi_t + f(u)\varphi_x)dx\, dt + \int_{-\infty}^\infty v^0(x)\varphi(0,x)dx$$

$$+ v_-(f'(u_-) - g(u_-)) < \delta_{\Lambda_-}, \varphi > + v_+(f'(u_+) - g(u_+)) < \delta_{\Lambda_+}, \varphi > = 0,$$

where $\Lambda_\pm = \{(t, f'(u_\pm)t)\}$; $\int_{-\infty}^\infty v^0(x)\varphi(0,x)dx = < v^0 \delta_{t=0}, \varphi >$ etc.

8.2   Weak continuous solutions for the scalar conservation laws

1.   To begin with we shall consider the scalar conservation law (8.5) equipped with the Cauchy data $u_0(x)$ s.t. $u_0 \in C^1(\mathbf{R}^1)$, u_0, $u_0' \in L^\infty(\mathbf{R}^1)$. We shall look for a classical solution $u \in C^1(\Omega) \cap C^0(\bar{\Omega})$ under the rather restrictive assumption $f'' > 0$ everywhere ($f \in C^2(\mathbf{R}^2)$).

As usual, the characteristics satisfy the ODE

$$\frac{dx}{dt} = f'(u(t, x(t))) \qquad (8.33)$$

$$x|_{t=0} = \alpha \in \mathbf{R}^1.$$

Thus, the characteristic $L : x = x(t, \alpha)$ (shortly $x(t)$) is smooth with respect to (t, α). Put $u|_L = u(t, x(t)) \in C^1$. Then $\frac{d}{dt}(u|_L) = \frac{\partial u}{\partial t} + \frac{\partial u}{\partial x}\frac{dx}{dt} = u_t + (f(u))_x = 0 \Rightarrow u|_L = const = u(0, x(0, \alpha)) = u(0, \alpha) = u_0(\alpha)$. Therefore, u is equal to a constant along any characteristic curve L.

Moreover, (8.33) implies that $\frac{dx}{dt} = f'(u_0(\alpha))$, $x|_{t=0} = \alpha \Rightarrow L : x = \alpha + tf'(u_0(\alpha))$, $u|_L = u_0(\alpha)$. The characteristics turn out to be straight lines in $\mathbf{R}^2$.

Consider now two characteristics L_1, L_2 and let they cross at a point $P \in \Omega$, $\Omega = \{t > 0\}$. Without loss of generality we shall assume that the characteristics L_1, L_2 correspond to the values of the parameter α : $\alpha_1 < \alpha_2$. Thus, $L_i : x = \alpha_i + tf'(u_0(\alpha_i))$, $i = 1, 2$. If L_1 crosses L_2 then $f'(u_0(\alpha_1)) \neq f'(u_0(\alpha_2))$ as otherwise $L_1 \parallel L_2$. Moreover, $f'' > 0 \Rightarrow f'$ is strictly monotonically increasing, i.e. $u_0(\alpha_1) \neq u_0(\alpha_2)$. The crossing point $P(t, x)$ is such that $t = \frac{\alpha_1 - \alpha_2}{f'(u_0(\alpha_2)) - f'(u_0(\alpha_1))}$ and consequently $t > 0 \Rightarrow f'(u_0(\alpha_2)) < f'(u_0(\alpha_1)) \iff u_0(\alpha_2) < u_0(\alpha_1)$. Then u is double valued, as $u(P) = u|_{L_1} = u|_{L_2} = u_0(\alpha_1) = u_0(\alpha_2)$.

Conclusion. Any two characteristics L_1, L_2 are not crossing for $t > 0$ if and only if the function $u_0(\alpha)$ is monotonically increasing. Geometrically, the characteristic L_2 is located to the right with respect to L_1. Certainly, L_2 is starting from α_2 and α_2 is located on the real axes $\overrightarrow{0x}$ to the right with respect to α_1. The solution u is then single valued, of course.

Evidently, $u(t, x) = u_0(\alpha)$ and in order to solve (8.5) we must invert the smooth mapping $\mathbf{R}^1_\alpha \to \mathbf{R}^1_x : x = \alpha + tf'(u_0(\alpha))$ depending smoothly on the parameter $t > 0$. Certainly, $\alpha \to \pm\infty \Rightarrow x \to \pm\infty$ as $u_0 \in L^\infty$. So the mapping is "onto". On the other hand, $\frac{\partial x}{\partial \alpha} = 1 + tf''(u_0(\alpha))u_0'(\alpha) \geq 1$ if $u_0' \geq 0$, i.e. if u_0 is monotonically increasing. The inverse mapping theorem gives us the existence of the inverse mapping $\alpha(t, x) \in C^1(\bar{\Omega}) \Rightarrow u(t, x) = u_0(\alpha(t, x))$ etc.

We have found a necessary and sufficient condition for the global classical solvability of (8.5) ($u \in C^1(t \geq 0)$), namely $u_0' \geq 0$ everywhere and we have constructed the unique solution u.

Suppose now that u_0 is not strictly monotonically increasing. Then we can not invert $x(t, \alpha)$ smoothly if $1 + t_0 f''(u_0(\alpha_0))u_0'(\alpha_0) = 0$ for some $t_0 > 0$, α_0, i.e. for $t_0 = \frac{1}{-\frac{d}{d\alpha} f'(u_0(\alpha_0))}$ ($u_0'(\alpha_0) < 0$).

Put $0 < T = \frac{1}{-\inf[\frac{d}{d\alpha} f'(u_0(\alpha_0))]}$. Then there exists a smooth solution $u \in C^1([0, T) \times \mathbf{R}^1_x)$. T is called life span time of u.

We shall discuss below the existence of continuous generalized (weak) solutions of the Cauchy problem (8.5) in the sense of Definition 8.1.

Remark 8.3. Suppose that $\Gamma : x = \gamma(t)$ is a smooth curve in $\Omega = \{t > 0\}$ starting from the $\overrightarrow{0x}$ - axes. Let Γ divide Ω into two open parts, $\Omega_\pm = \{(t,x) : \pm(x - \gamma(t)) > 0\}$ and $u \in C^1(\bar{\Omega}_+) \cap C^1(\bar{\Omega}_-)$ is a weak solution of (8.5). Suppose that ∇u has a jump at the point $P_0 = (t_0, \gamma(t_0)) \in \Omega$. Then $\gamma(t)$ satisfies for $|t - t_0| \ll 1$ the ODE $\gamma'(t) = f'(u(t, \gamma(t)))$, i.e. Γ is a characteristic for the scalar conservation law equation (8.5). Thus, jumps propagate along the characteristics only.

To simplify the things we introduce the notations $u_+ = u|_{\bar{\Omega}_+}$, $u_- = u|_{\bar{\Omega}_-}$. As we know, $u_\pm$ satisfy in classical sense (8.5) in $\Omega_\pm \Rightarrow u_\pm$ satisfy in classical sense (8.5) in $\bar{\Omega}_\pm$. Having in mind that $u_+|_\Gamma = u_-|_\Gamma = u|_\Gamma$ we get:

$$u_+(t, \gamma(t)) = u_-(t, \gamma(t)) \Rightarrow \frac{\partial u_+}{\partial t}(t, \gamma(t)) + \frac{\partial u_+}{\partial x}(t, \gamma(t))\gamma'$$

$$= \frac{\partial u_-}{\partial t}(t, \gamma(t)) + \frac{\partial u_-}{\partial x}(t, \gamma(t))\gamma', \forall t \geq 0.$$

Thus,

$$\frac{\partial u_+}{\partial x}(t, \gamma)\gamma' - f'(u_+(t, \gamma))\frac{\partial u_+}{\partial x} = \frac{\partial u_-}{\partial x}(t, \gamma)\gamma' - f'(u_-(t, \gamma))\frac{\partial u_-}{\partial x}.$$

The equality $u_+|_\Gamma = u_-|_\Gamma = u|_\Gamma$ implies that $(\frac{\partial u_+}{\partial x}(t, \gamma) - \frac{\partial u_-}{\partial x}(t, \gamma))(\gamma' - f'(u(t, \gamma))) = 0$. Then $\frac{\partial u_+}{\partial x}(t_0, \gamma(t_0)) \neq \frac{\partial u_-}{\partial x}(t_0, \gamma(t_0))$ gives us that $\gamma'(t) = f'(u(t, \gamma(t)))$ for $|t - t_0| \ll 1$.

Corollary 8.2. *The finite jump type singularities of* ∇u *of the continuous generalized (weak) solution of (8.5) are propagating along the characteristics of (8.5).*

2. From now on we shall follow with unsignificant changes and small simplifications the proof of Theorem 1 from [40] (see also [39]).

Our aim is to prove that if u is a continuous weak solution of (8.5) (see Definition 8.3) then $u = const$ along any characteristic of (8.5). Moreover, as in the classical $C^1(t \geq 0)$ case a unique characteristic emanates from each point $(\bar{t}, \bar{x}) \in [0, T) \times \mathbf{R}^1_x$ and its graph is a straight line in $\mathbf{R}^2_{t,x}$ with slope $f'(u(\bar{t}, \bar{x}))$. We point out that we can assume that $u_0 \in L^\infty_{loc}$, $u \in L^\infty_{loc}(t \geq 0) \Rightarrow f(u) \in L^\infty_{loc}(t \geq 0)$.

Remark 8.4. Let $u \in C^1(\Omega) \cap C^0(\bar{\Omega})$ be a classical solution of (8.5) with $f''(u) > 0$, $\forall u \in \mathbf{R}^1$. Let $L : x = \gamma(t) \in C^1$ be a characteristic of (8.5), i.e. $\gamma' = f'(u(t, \gamma(t)))$. We shall prove in another way that γ is a linear

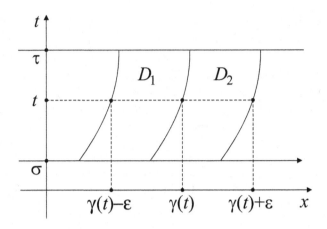

Fig. 8.3

function. To do this fix any $0 < \sigma < \tau$ and $\varepsilon > 0$ and define the open domains $D_1 = \{(t,x) : \sigma < t < \tau, \gamma(t) - \varepsilon < x < \gamma(t)\}$, $D_2 = \{(t,x) : \sigma < t < \tau, \gamma(t) < x < \gamma(t) + \varepsilon\}$ (see Fig. 8.3), $u \in C^1(\bar{D}_{1,2})$

Then

$$0 = \iint_{D_1} \{u_t + (f(u))_x\} dx\, dt = I_1 + I_2,$$

where

$$I_2 = \int_\sigma^\tau \left\{ \int_{\gamma(t)-\varepsilon}^{\gamma(t)} (f(u))_x dx \right\} dt = \int_\sigma^\tau \{f(u(t,\gamma(t))) - f(u(t,\gamma(t) - \varepsilon))\} dt,$$

$$I_1 = \int_\sigma^\tau \left\{ \int_{\gamma(t)-\varepsilon}^{\gamma(t)} \frac{\partial u}{\partial t}(t,x) dx \right\} dt.$$

put $F(t,x) = \int_{\gamma(t)-\varepsilon}^{\gamma(t)} u(t,x) dx$. Evidently, $\frac{\partial F}{\partial t} = \int_{\gamma(t)-\varepsilon}^{\gamma(t)} \frac{\partial u}{\partial t}(t,x) dx + \gamma'(t)u(t,\gamma(t)) - u(t,\gamma(t) - \varepsilon)\gamma'(t)$ and therefore

$$I_1 = \int_\sigma^\tau \frac{\partial F}{\partial t}(t,x) dt - \int_\sigma^\tau \gamma'(t)\{u(t,x(t)) - u(t,\gamma(t) - \varepsilon)\} dt$$

$$= \int_{\gamma(\tau)-\varepsilon}^{\gamma(\tau)} u(\tau,x) dx - \int_{\gamma(\sigma)-\varepsilon}^{\gamma(\sigma)} u(\sigma,x) d\sigma$$

$$- \int_\sigma^\tau \gamma'(t)\{u(t,\gamma(t)) - u(t,\gamma(t) - \varepsilon)\} dt$$

Consequently,

$$\int_{\gamma(\tau)-\varepsilon}^{\gamma(\tau)} u(\tau,x)dx - \int_{\gamma(\sigma)-\varepsilon}^{\gamma(\sigma)} u(\sigma,x)dx \qquad (8.34)$$

$$= \int_\sigma^\tau \{f(u(t,\gamma(t)-\varepsilon)) - f(u(t,\gamma(t)))$$

$$- \gamma'(t)(u(t,\gamma(t)-\varepsilon) - u(t,\gamma(t)))\}dt.$$

On the other hand, Taylor's formula gives us that

$$f(u(t,\gamma(t)-\varepsilon)) - f(u(t,\gamma(t))) = (u(t,\gamma(t)-\varepsilon) - u(t,\gamma(t)))f'(u(t,\gamma(t)))$$

$$+\frac{1}{2}(u(t,\gamma(t)-\varepsilon) - u(t,\gamma(t)))^2 f''(\lambda), \gamma' = f'(u(t,\gamma)),$$

$\lambda \in (u(t,\gamma(t)), u(t,\gamma(t)-\varepsilon))$. Thus (8.34) implies

$$\int_{\gamma(\tau)-\varepsilon}^{\gamma(\tau)} u(\tau,x)dx - \int_{\gamma(\sigma)-\varepsilon}^{\gamma(\sigma)} u(\sigma,x)dx \geq 0. \qquad (8.35)$$

Dividing (8.35) by $\varepsilon > 0$ and letting $\varepsilon \to 0$ we get

$$u(\tau,\gamma(\tau)) - u(\sigma,\gamma(\sigma)) \geq 0. \qquad (8.36)$$

In a similar way

$$0 = \iint_{D_2} \{u_t + (f(u))_x\}dx\,dt = II_1 + II_2,$$

where $II_2 = \int_\sigma^\tau \{\int_{\gamma(t)}^{\gamma(t)+\varepsilon}(f(u))_x dx\}dt$,

$$II_1 = \int_\sigma^\tau \left\{\int_{\gamma(t)}^{\gamma(t)+\varepsilon} \frac{\partial u}{\partial t}(t,x)dx\right\}dt.$$

Introducing the function $G(t,x) = \int_{\gamma(t)}^{\gamma(t)+\varepsilon} u(t,x)dx$ we obtain easily that

$$\int_{\gamma(\tau)}^{\gamma(\tau)+\varepsilon} u(\tau,x)dx - \int_{\gamma(\sigma)}^{\gamma(\sigma)+\varepsilon} u(\sigma,x)dx \qquad (8.37)$$

$$= -\int_\sigma^\tau \{f(u(t,\gamma(t)+\varepsilon)) - f(u(t,\gamma(t))) - \gamma'(t)(u(t,\gamma(t)+\varepsilon) - u(t,\gamma(t)))\}dt.$$

Applying again Taylor's formula we get that

$$\int_{\gamma(\tau)}^{\gamma(\tau)+\varepsilon} u(\tau,x) - \int_{\gamma(\sigma)}^{\gamma(\sigma)+\varepsilon} u(\sigma,x)dx \leq 0 \qquad (8.38)$$

and therefore

$$u(\tau, \gamma(\tau)) - u(\sigma, \gamma(\sigma)) \le 0. \tag{8.39}$$

Combining (8.36) -(8.39) we conclude that $u(\tau, \gamma(\tau)) = u(\sigma, \gamma(\sigma))$, τ, σ being arbitrary. If $L : \gamma = \gamma(s)$, $0 < s \le T$ is a characteristic of (8.5) we have proved again that $u|_L = const = u(P_0)$, where P_0 is the crossing point of L with the $\overrightarrow{0x}$ axes (by continuity as $u \in C^0(t \ge 0)$). Therefore, $\gamma'(t) = f'(u(P_0)) \Rightarrow \gamma(t) = tf'(u(P_0)) + C_1$, i.e. $L : x = \gamma(t)$ is a straight line in the plane.

  3. We give below the following interesting result.

Proposition 8.3. *Let $L : x = \gamma(t)$, $0 < t < T$ be any characteristic associated with the continuous weak solution u of (8.5). Then $u|_L = \bar{u} = const$. More precisely, the unique characteristic L starting from the point $(\bar{t}, \bar{x}) \in [0, T) \times \mathbf{R}_x^1$ is a straight line with slope $f'(u(\bar{t}, \bar{x}))$.*

Proof. The main idea is to verify that the identities (8.34), (8.37) remain true in the case of continuous weak solutions of (8.5). Then (8.34), (8.37) imply Proposition 8.3 as in Remark 8.4 (smooth case). To simplify the proof and to fix the ideas we shall investigate the generalized solution u in $t > 0$ only. Thus the function $u \in C^0(\Omega)$ satisfies the integral identity

$$\int_0^T \int_{-\infty}^\infty (u\varphi_t + f(u)\varphi_x)dt\,dx = 0, \forall \varphi \in C_0^\infty(\Omega). \tag{8.40}$$

According to [110] the identity (8.40) holds for each compactly supported function $\varphi \in H_1^\infty = \{g : g \in L^\infty, \nabla g \in L^\infty\}$. Here ∇g is the (distribution) gradient of g. According to Chapter V, Section 6 of [110] $g \in H_1^\infty \iff g \in L^\infty, sup_{x,x'} \frac{|g(x)-g(x')|}{|x-x'|} < \infty$.

  In other words we can put in (8.40) the Lipschitz function $\varphi(t, x) = h(t)\psi(t, x)$ instead of $\varphi \in C_0^\infty$, where

$$\psi(t,x) = \begin{cases} 0, & 0 < t < T, \quad -\infty < x \le \gamma(t) - \varepsilon - \delta \\ \frac{1}{\delta}(x - \gamma(t) + \varepsilon + \delta), & 0 < t < T, \quad \gamma(t) - \varepsilon - \delta < x \le \gamma(t) - \varepsilon \\ 1, & 0 < t < T, \quad \gamma(t) - \varepsilon < x \le \gamma(t) \\ \frac{1}{\delta}(-x + \gamma(t) + \delta), & 0 < t < T, \quad \gamma(t) < x \le \gamma(t) + \delta \\ 0, & 0 < t < T, \quad \gamma(t) + \delta < x < \infty, \end{cases}$$

$$h(t) = \begin{cases} 0, & 0 < t \le \sigma - \delta \\ \frac{1}{\delta}(t - \sigma + \delta), & \sigma - \delta < t \le \sigma \\ 1, & \sigma < t \le \tau \\ \frac{1}{\delta}(-t + \tau + \delta), & \tau < t \le \tau + \delta \\ 0, & \tau + \delta < t < T, \end{cases}$$

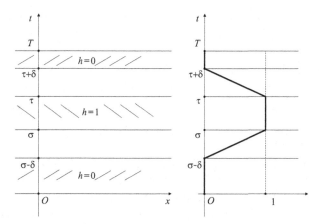

Fig. 8.4

with $\delta > 0$ small. As we know, $\gamma(t)$ is a smooth function and therefore ψ and h are piecewise smooth compactly supported and continuous functions, i.e. $\varphi = \psi h \in H_1^\infty$ (see Fig. 8.4).

Certainly, $\varphi_t = \psi_t h + \psi h_t$, $\varphi_x = \psi_x h$ and $\int_0^T (.)dt = \int_{\sigma-\delta}^\sigma (.)dt + \int_\sigma^\tau (.)dt + \int_\tau^{\tau+\delta} (.)dt$, while $\int_{-\infty}^\infty (.)dx = \int_{\gamma(t)-\varepsilon-\delta}^{\gamma(t)-\varepsilon} (.)dx + \int_{\gamma(t)-\varepsilon}^{\gamma(t)} (.)dx + \int_{\gamma(t)}^{\gamma(t)+\delta} (.)dx$.

Therefore, we must investigate several integrals, letting $\delta \to 0$. Obviously, the double integral over $\{(\sigma < t < \tau) \times (\gamma(t) - \varepsilon < x < \gamma(t)\}$ is 0.

The double integral (8.40) produces two types of integrals: the first ones having denominator δ^2 and the second ones -denominator δ.

Below we shall study 3 examples only.

$$I = \int_{\sigma-\delta}^\sigma \int_{\gamma(t)-\varepsilon-\delta}^{\gamma(t)-\varepsilon} (u\varphi_t + f(u)\varphi_x)dx\,dt$$

$$= \frac{1}{\delta^2} \int_{\sigma-\delta}^\sigma \int_{\gamma(t)-\varepsilon-\delta}^{\gamma(t)-\varepsilon} -\gamma'(t)(t - \sigma + \delta)u(t,x)dt\,dx$$

$$+ \frac{1}{\delta^2} \int_{\sigma-\delta}^\sigma \int_{\gamma(t)-\varepsilon-\delta}^{\gamma(t)-\varepsilon} u(t,x)(x - \gamma(t) + \varepsilon + \delta)dt\,dx$$

$$+ \frac{1}{\delta^2} \int_{\sigma-\delta}^\sigma \int_{\gamma(t)-\varepsilon-\delta}^{\gamma(t)-\varepsilon} f(u(t,x))(t - \sigma + \delta)dx\,dt \equiv I_1 + I_2 + I_3.$$

Put $D_3 = \{(t,x) : \sigma - \delta < t < \sigma, \gamma(t) - \varepsilon - \delta < x < \gamma(t) - \varepsilon\}$.

Then D_3 is a bounded domain and $t - \sigma + \delta \geq 0$, $\gamma' \in C^0$, $x - \gamma(t) + \varepsilon + \delta \geq 0$ there, $u \in C^0(\bar{D}_3)$. Therefore,

$$|I_1| \leq \frac{C}{\delta^2} \int_{\sigma-\delta}^{\sigma} \left\{ \int_{\gamma(t)-\varepsilon-\delta}^{\gamma(t)-\varepsilon} (t - \sigma + \delta) dx \right\} dt = \frac{C}{\delta^2} \delta \int_{\sigma-\delta}^{\sigma} (t - \sigma + \delta) dt = \frac{C}{2} \delta,$$

$C = const > 0$; $0 \leq x - \gamma(t) + \varepsilon + \delta \leq \delta$ implies that $|I_2| \leq \frac{C}{\delta^2} \delta \int_{\sigma-\delta}^{\sigma} \{ \int_{\gamma(t)-\varepsilon-\delta}^{\gamma(t)-\varepsilon} 1. dx \} dt = C\delta$. Evidently, $|I_3| \leq C\delta$. This way we conclude that $lim_{\delta \to 0} I = 0$.

In a similar way all integrals having denominator δ^2 tend to 0 for $\delta \to 0$.

Denote $D_4 = \{(t, x) : \sigma - \delta < t < \sigma, \gamma(t) - \varepsilon < x < \gamma(t)\}$ and consider the corresponding integral over D_4, where $h(t) = \frac{1}{\delta}(t - \sigma + \delta)$, $\psi(t, x) = 1 : II = \iint_{D_4} (\varphi_t u + f(u)\varphi_x) dx \, dt = \frac{1}{\delta} \int_{\sigma-\delta}^{\sigma} \{ \int_{\gamma(t)-\varepsilon}^{\gamma(t)} u(t, x) dx \} dt$

$\Rightarrow lim_{\delta \to 0} II = \int_{\gamma(\sigma)-\varepsilon}^{\gamma(\sigma)} u(\sigma, x) dx$ according to L'Hospital rule.

Let $III = \int_{\sigma}^{\tau} \int_{\gamma(t)}^{\gamma(t)+\delta} (\varphi_t u + f(u)\varphi_x) dx \, dt = \frac{1}{\delta} \int_{\gamma(t)}^{\gamma(t)+\delta} \{ \int_{\sigma}^{\tau} \gamma' u(t, x) dt \} dx - \frac{1}{\delta} \int_{\gamma(t)}^{\gamma(t)+\delta} \{ \int_{\sigma}^{\tau} f(u(t, x)) dt \} dx$.

Applying again L'Hospital rule to the integrals participating in the right-hand side of III we get:

$$lim_{\delta \to 0} III = \int_{\sigma}^{\tau} \gamma'(t) u(t, \gamma(t)) dt - \int_{\sigma}^{\tau} f(u(t, \gamma(t))) dt.$$

We complete the proof of the identity (8.34) and of inequalities (8.35) and (8.36) in an obvious way. In a similar way we verify (8.37) and as a consequence we obtain (39). The proof of Proposition 8.3 ends the same way as the one of Remark 8.4.

4. We propose below several comments on Proposition 8.3. The characteristic $L : x = \gamma(t)$, $\bar{x} = \gamma(\bar{t})$, $\bar{t} > 0$ satisfies the ODE $\gamma'(t) = f'(u(t, \gamma(t)))$, the right-hand side being only continuous as $u \in C(t > 0)$ and $f' \in C^1$. Peano existence theorem gives us the local existence of a solution $\gamma \in C^1$ but does not guarantee its uniqueness. Eventually, there could exist several solutions. Proposition 8.3 verifies the uniqueness of L. Moreover, L turns out to be a straight line passing through $(\bar{t}, \bar{x})$ and with a slope $f'(u(\bar{t}, \bar{x}))$. This is not an obvious result, of course, if we start from Definition 8.1 (Def.8.3) and the integral identity (8.40).

Corollary 8.3. *Under the assumptions of Proposition 8.3 for any $x < y$ and $0 < t < T$ the following inequality holds:*

$$\frac{-1}{T - t} \leq \frac{f'(u(t, y)) - f'(u(t, x))}{y - x} \leq \frac{1}{t}. \tag{8.41}$$

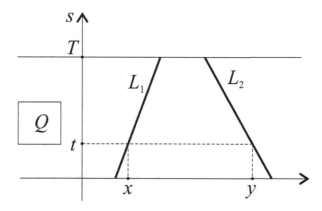

Fig. 8.5

Proof. Consider the characteristics $L_1 : z = x + (s-t)f'(u(t,x))$, $L_2 : z = y + (s-t)f'(u(t,y))$ starting for $s = t$ from the points $(t,x) \in L_1$, respectively $(t,y) \in L_2$. As we know these two straight lines cannot cross in the strip $\{0 < \tau < T\}$ as the continuous solution $u(t,x)$ is single-valued there. Therefore, L_1 is located to the left with respect to L_2 (see Fig. 8.5).

Consequently, $x + (s-t)f'(u(t,x)) < y + (s-t)f'(u(t,y))$ for $0 < s < T$.

Letting $s \to 0$ and $s \to T$ in the previous inequality we obtain (8.41). In our case $f'' > 0$ everywhere and therefore in each closed bounded rectangle $Q \subset \{0 < \tau < T\}$ the solution u is Lipschitz with respect to x : $|u(t,y) - u(t,x)| \le C(Q)|x-y|$. If $const = a < f'' \in L^\infty$, $a > 0$, then $-\frac{c_2}{T-t} \le \frac{u(t,y)-u(t,x)}{y-x} \le \frac{c_1}{t}$, for each $x < y$ and each $0 < t < T$, c_1, $c_2 > 0$ being constants. This way we come to a regularity result for the continuous weak solution u of (8.5) (interior regularity result in the strip $\{0 < t < T\}$ as (8.40) holds). It turns out to be Lipschitz continuous with respect to x. Thus for each $T > t > 0$ fixed the continuous function $u(t,x)$ is differentiable (in a classic sense) for almost all $x \in \mathbf{R}^1$ [47].

Example 1. Consider the Cauchy problem $u_t + uu_x = 0$, $t \ge 0$, $u|_{t=0} = u_0(x) = \begin{cases} 0, & x \le 0 \\ x, & x \ge 0, \end{cases}$ i.e. $u_0 \in C^0(\mathbf{R}^1)$ and u_0 is a Lipschitz function. Then by the method of characteristics one can construct the weak solution $u(t,x) = \begin{cases} 0, & x \le 0 \\ \frac{x}{t+1}, & x > 0 \end{cases}$ of the initial value problem in the sense of Definition 8.1 (i.e. $u \in L^1_{loc}(t \ge 0)$). Evidently, $u \in C^0(t \ge 0)$ and moreover, the finite jump of $u_0'(x)$ at $x = 0$ propagates along the characteristic

$x = 0$, $t > 0$ as a finite jump of u_x; $u(t,x)$ is a Lipschitz function with respect to x.

Example 2. We shall deal with the Cauchy problem $u_t + uu_x = 0$, $u|_{t=0} = u_0(x) = \begin{cases} 0, & x \le 0 \\ \sqrt{x}, & x > 0, \end{cases}$ $t \ge 0$. Certainly, $u_0 \in C^0(\mathbf{R}^1)$ and $u_0(x)$ is Hölder continuous with exponent $1/2$. By the method of characteristics we conclude that the weak solution

$$u = \begin{cases} 0, & x \le 0 \\ \frac{1}{2}(\sqrt{t^2 + 4x} - t), & x > 0 \end{cases} \in C^0(t \ge 0); u_x = \begin{cases} 0, & x \le 0 \\ \frac{1}{\sqrt{t^2+4x}}, & x > 0. \end{cases}$$

Obviously, $u = \frac{2x}{t+\sqrt{t^2+4x}}$ for $x > 0$ and therefore for each $t > 0$ fixed $u(t,x)$ is Lipschitz continuous with respect to x. The derivative u_x has an infinite jump at the origin, while $u_x(t,0)$ for $t > 0$ and fixed has a finite jump $|\frac{1}{t}|$. Therefore, starting with infinite jump type singularity of u_0' at 0 we have propagation of singularities of the type finite jump of u_x along the characteristic $L : x = 0, t > 0$. The function $u_x \in C^\infty(t \ge 0, x > 0)$ and $u_x \in C^\infty(x \le 0, t > 0)$. Therefore, Hölder type singularities of u_0 transforms in Lipschitz type singularity of u which propagates along L for $t > 0$.

5. **Concluding remarks.** Let $u \in C^0(t \ge 0)$ be a weak solution of (8.5) such that $u_0(x) = u|_{t=0} \in L^\infty(\mathbf{R}^1) \cap C^0(\mathbf{R}^1)$ and u_0 is monotonically increasing; $f'' > 0$. We shall propose here a procedure of constructing of u. It repeats the considerations from the begining of this section where a $C^1(t \ge 0)$ solution u of (8.5) was constructed. The approach relies heavily on Proposition 8.3 as the relation $u|_L = \bar{u}$, L being arbitrary characteristic, holds both in classical (C^1) and generalized (C^0) cases. Thus, $u|_L = u_0(\alpha)$, where the characteristic $L : x = \alpha + tf'(u_0(\alpha))$, $t \ge 0$, is a straight line in the plane $\mathbf{R}^2_{t,x}$ starting for $t = 0$ by the point $\alpha \in \mathbf{R}^1$. Obviously, $x = x(t,\alpha)$ is a strictly monotonically increasing continuous function of $\alpha \in \mathbf{R}^1$ for each fixed $t \ge 0$ (see Fig. 8.6).

More precisely, for each fixed $t \ge 0$ the continuous mapping $x(t,\alpha) :$ $\mathbf{R}^1_\alpha \to \mathbf{R}^1_x$ is "onto" as $u_0 \in L^\infty$. Moreover, $x(t,\alpha)$ is invertible: $\exists ! \, \alpha(t,x) :$ $\mathbf{R}^1_x \to \mathbf{R}^1_\alpha$. Thus, by the definition of α

$$x = \alpha(t,x) + tf'(u_0(\alpha(t,x))) \tag{8.42}$$

For every $t \ge 0$ fixed $\alpha(t,x)$ is continuous and strictly monotonically increasing function of x. Our main problem is whether $\alpha(t,x) \in C^0(t \ge 0)$. If so, $u(t,x) = u_0(\alpha(t,x))$ etc.

Suppose now that the sequence $\{(t_n, x_n)\} \to (t_0, x_0)$, $x_0 \in \mathbf{R}^1$, $t_0 \ge 0$. According to (8.42) $x_n = \alpha(t_n, x_n) + t_n f'(u_0(\alpha(t_n, x_n)))$. There are two

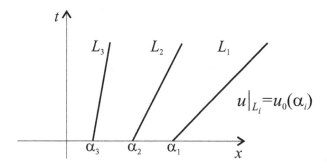

Fig. 8.6

possibilities for the sequence $\{\alpha(t_n, x_n)\}$: a) $\{\alpha(t_n, x_n)\}$ is unbounded and b) $\{\alpha(t_n, x_n)\}$ is bounded. In the case a) there exists a subsequence $\{\alpha(t_{n_k}, x_{n_k})\} \to +\infty(-\infty)$. Having in mind that $lim_{z_n \to \pm\infty} u_0(z) = u_0(\pm\infty) \in \mathbf{R}^1$ exist as u_0 is a bounded monotonically increasing function we obtain from (8.42) that $lim x_{n_k} = +\infty(-\infty)$ - contradiction with $lim_{k \to \infty} x_{n_k} = x_0 \in \mathbf{R}^1$. In the case b) let $l \in \mathbf{R}^1$ be any accumulation point of the bounded sequence $\{\alpha(t_n, x_n)\}$. Then one can find such a subsequence that $\{\alpha(t_{n_k}, x_{n_k})\} \to l$. Applying again (8.42) we get that there exists $x_0 = lim_{k \to \infty} x_{n_k} = l + t_0 f'(u_0(l)) \Rightarrow l = \alpha(t_0, x_0)$. Consequently, there exists a unique accumulation point of $\{\alpha(t_n, x_n)\} \Rightarrow \{\alpha(t_n, x_n)\} \to \alpha(t_0, x_0)$ for $\{(t_n, x_n)\} \to (t_0, x_0)$. This way we have shown that $\alpha \in C^0(t \geq 0)$. So the generalized Cauchy problem has a unique solution in the class of the continuous functions $C^0(t \geq 0)$ and for $u_0 \in L^\infty$ monotonically increasing and continuous.

Having in mind Lebesgue theorem on the differentiability of monotonic functions almost everywhere (a.e.) one can easily prove that the above constructed solution $u \in C(t \geq 0)$ is differentiable a.e. in the upper half plane $\{t \geq 0\}$. To do this two cases will be studied separately.

Case 1. Fix $\bar{t} > 0$. Then the function $u(\bar{t}, x) = u_0(\alpha(\bar{t}, x))$ is monotonically increasing function of x as a composition of two monotonically increasing functions and consequently $u(\bar{t}, x)$ is differentiable a.e. with respect to $x \in \mathbf{R}^1$.

Case 2. Fix $\bar{x}$ and consider $u(t, \bar{x}) = u_0(\alpha(t, \bar{x}))$. Let $t_1 > t_2 > 0$ $(0 < t_1 < t_2)$. The characteristics $L_1 : x = \alpha_1 + t f'(u_0(\alpha_1))$ and $L_2 : x = \alpha_2 + t f'(u_0(\alpha_2))$ are passing for $t = t_1$, respectively $t = t_2$ through the point $\bar{x}$, i.e. $\bar{x} = \alpha_1 + t_1 f'(u_0(\alpha_1))$, $\bar{x} = \alpha_2 + t_2 f'(u_0(\alpha_2))$. Applying (8.42) we get that $\alpha_1 = \alpha(t_1, \bar{x})$, $\alpha_2 = \alpha(t_2, \bar{x})$ and therefore

$$\alpha_2 - \alpha_1 = \alpha(t_2, \bar{x}) - \alpha(t_1, \bar{x}) = t_1 f'(u_0(\alpha_1)) - t_2 f'(u_0(\alpha_2)) \qquad (8.43)$$

$$= (t_1 - t_2) f'(u_0(\alpha_1)) - t_2 (f'(u_0(\alpha_2)) - f'(u_0(\alpha_1)))$$

$$= (t_1 - t_2) f'(u_0(\alpha_1)) - t_2 (u_0(\alpha_2) - u_0(\alpha_1)) f''(\lambda),$$

where $\lambda(t_1, t_2, \bar{x}) = u_0(\alpha_1) + \Theta(u_0(\alpha_2) - u_0(\alpha_1))$, $0 < \Theta < 1$.
Put $\Delta_{t_1,t_2}\alpha = \frac{\alpha_2 - \alpha_1}{t_2 - t_1} = \frac{\alpha(t_2,\bar{x}) - \alpha(t_1,\bar{x})}{t_2 - t_1}$.
From (8.43) we get that

$$\Delta_{t_1,t_2}\alpha = -f'(u_0(\alpha_1)) - t_2 \Delta_{t_1,t_2}\alpha \frac{u_0(\alpha_2) - u_0(\alpha_1)}{\alpha_2 - \alpha_1} f''(\lambda),$$

i.e.

$$\Delta_{t_1,t_2}\alpha \left(1 + t_2 \frac{u_0(\alpha_2) - u_0(\alpha_1)}{\alpha_2 - \alpha_1} f''(\lambda)\right) = -f'(u_0(\alpha_1)).$$

The expression in the brackets is positive, of course. Fix now t_1 and suppose that $u_0'(\alpha_1)$ exists (almost all real numbers have this property). Let then $t_2 \to t_1 \Rightarrow \alpha_2 = \alpha(t_2, \bar{x}) \to \alpha(t_1, \bar{x}) = \alpha_1 \Rightarrow \frac{u_0(\alpha_2) - u_0(\alpha_1)}{\alpha_2 - \alpha_1} \to u_0'(\alpha_1) \geq 0$ and $f''(\lambda) \to f''(u_0(\alpha_1))$.

Consequently, $\frac{\partial \alpha}{\partial t}(t_1, \bar{x})$ exists and $\frac{\partial \alpha}{\partial t}(t_1, \bar{x}) = -f'(u_0(\alpha_1)) \,/\, 1 + t_1 u_0'(\alpha_1) f''(u_0(\alpha_1)))$, $\alpha_1 = \alpha(t_1, \bar{x})$.

So the global regularity result in $\{t \geq 0\}$ is shown.

Remark 8.5. The generalized solution u can be studied also when u is not obliged to be continuous for $t \geq 0$. To orientate in the situation we shall study the initial value problem $u_t + (f(u))_x = 0$, $u|_{t=0} = u_0 \in C^2(\mathbf{R}^2)$ $\Rightarrow u_x|_{t=0} = u_0'(x)$, $u_0, u_0' \in L^\infty$ and we shall suppose that $u \in C^3_{loc}(T > t \geq 0)$, $f(u) \in C^2$, $f''(u) \geq 0$ (see (8.5), (8.33)). Define now $z = u_x$ and differentiate the equation with respect to x. Thus

$$0 = z_t + f'(u)z_x + f''(u)z^2 \geq z_t + f'(u)z_x.$$

Consider then the characteristic $\Gamma : x = \gamma(t)$, $\gamma' = f'(u(t, \gamma(t)))$ passing for $t = 0$ through the point $\alpha \in \mathbf{R}^1$. The above inequality takes the following form along Γ:

$$z = z(t, \gamma(t)) : 0 \geq z_t + f'(u)z_x = \frac{dz}{dt}.$$

Therefore, for each $(t, x) \in \{0 \leq t < T, x \in \mathbf{R}^1\}$, where the classical smooth solution u exists we have that

$$z(t, \gamma(t)) = u_x(t, x) \leq z(0, \gamma(0)) = z(0, \alpha) = u_0'(\alpha) \leq \sup u_0' = K.$$

If T is the life span of u the derivative $u_x(t,x)$ could not be defined for $t = T$ and for some x. Therefore, instead of $u_x(t,x) \leq K$ we shall write

$$(*) \qquad \frac{u(t,x_2) - u(t,x_1)}{x_2 - x_1} \leq K, \forall x_1, x_2.$$

$(*)$ is called Oleinik admissible condition for the existence of u (see [88]). Fix $x = x^*$, $t = T$ and assume that u is not continuous at (T, x^*) (say, has a finite jump there). From $(*)$ it follows immediately that for

$$(**) \qquad x_2 \to x^* + 0, x_1 \to x^* - 0 \Rightarrow u(t, x^* + 0) < u(t, x^* - 0).$$

The condition $(**)$ is called admissibility condition at each finite jump point of the weak solution u, u being of the class of piecewise smooth functions.

Example 3. Consider the piecewise smooth (constant) function

$$u_\delta(t,x) = \begin{cases} 0, & x < -\delta t \\ -\delta, & -\delta t < x < 0 \\ \delta, & 0 < x < \delta t \\ 0, & x > \delta t \end{cases}, \delta > 0.$$ It satisfies in weak sense (see Definition 8.1) the Cauchy problem $u_t + 2uu_x = 0$, $u|_{t=0} = 0$. In fact, u_δ verifies outside $x = \mp \delta t$, $t > 0$, $x = 0$ the equation and the Rankine-Hugoniot conditions are satisfied along $x = \pm \delta t$, $t > 0$ and $x = 0$. Therefore, the problem possesses infinitely many weak solutions, including $u \equiv 0$. On the other hand, $(**)$ is not satisfied along $x = 0$, $t > 0$ as $u_+|_{x=0} = \delta$, while $u_-|_{x=0} = -\delta$. The only admissible solution turns out to be $u \equiv 0$. Details on the subject could be found in [88], [69], [71], [39], etc. Roughly speaking, conditions of the type $(**)$ guarantee the existence of unique weak solution.

Chapter 9

Microlocal approach in studying the propagation of nonlinear waves

9.1 Propagation of jump type discontinuities in $\mathbf{R}^2_{t,x}$

In the investigation of the propagation of singularities of the solutions to semilinear hyperbolic (non-strictly hyperbolic) equations and systems, new effects in comparison with the linear case can appear. The interaction of the singularities propagating along several characteristics crossing at some point (surface) could give rise of new singularities - usually weaker than the initial ones-propagating along the outgoing characteristics starting from that point (surface). The propagation of finite jump discontinuities of weakly hyperbolic systems in the case of one space variable was considered in [98], [99], [59] and others. To illustrate the situation we propose the following example [94]: Consider the weakly hyperbolic system in $\mathbf{R}^2$:

$$
\begin{vmatrix}
Xu = (\partial_t + pt^{p-1}\partial_x)u = 0 \\
\partial_t v = u \\
Dz = (\partial_t + \partial_x)z = uv,
\end{vmatrix}
$$

p being an even integer, $p > 0$ equipped with initial data $u_0(x)$, $v_0(x) = z_0(x) = 0$ prescribed for $t = -T < 0$. Denote by C_1, C_2, C_3 the characteristics of the vector fields X, ∂_t, D passing through the origin, namely $x = t^p$, $x = 0$, $x = t$. We shall study the following different cases:

a) $u_0 = \begin{cases} x, & x \le 0 \\ 2x, & x > 0 \end{cases}$, b) $u_0 = \begin{cases} x, & x \le 0 \\ 0, & x > 0 \end{cases}$, c) $u_0 = \begin{cases} 0, & x \le 0 \\ x, & x > 0 \end{cases}$,

d) $u_0 = \begin{cases} 1, & x \le 0 \\ 1+x, & x > 0 \end{cases}$.

Evidently, u_0 is continuous, while u_0' has a finite jump at $x = 0$.

One can check (see [94]) that with $C_j^+ = C_j \cap \{x > 0\}$, $j = 1, 2, 3$ we have: a,b) $z \in C^{3p}$ in a neighbourhood of C_3^+ contained inside the parabola

C_1, while some transversal with respect to C_3 derivative of $z(t,x)$ of order $3p + 1$ has a finite jump discontinuity at C_3^+.

c) $z \in C^\infty$ in a neighbourhood of C_3^+ located inside C_1.

d) $z \in C^{2p}$ near C_3^+ and the transversal derivatives of $z : \partial_x^{2p+1} z$, $\partial_t^{2p+1} z$ have finite jumps along C_3^+ inside C_1.

In other words, finite jump discontinuities of the Cauchy data give rise to new created finite jump discontinuities along C_3^+. The new singularities are weaker than the initial ones. Certainly, in our case the space variable $x \in \mathbf{R}^1$.

9.2 Creation of logarithmic singularities in $\mathbf{R}^3_{t,x_1,x_2}$

1. In this Chapter we shall give a model example of first order (non)strictly hyperbolic semilinear system in $\mathbf{R}^3_{t,x_1,x_2}$ discussing the singularities of its solutions. The initial data are assumed to have finite jump type discontinuities along two characteristic surfaces Σ_1, Σ_2 which cross transversally along the straight line $\Gamma = \Sigma_1 \cap \Sigma_2$. We are interested in the production of logarithmic singularities from the interaction of these piecewise smooth waves along Γ.

We give here some notations and we remind the reader of several notions. Thus, $\Box_c = \partial_t^2 - c^2 \triangle_x$, $c > 0$ is the wave operator in $\mathbf{R}^1_t \times \mathbf{R}^2_x$ and by $K_c^-(z_0)$ we denote the backward characteristic open cone with vertex at $z_0 = (t_0, x_0)$, $t_0 > 0$, i.e. $K_c^-(z_0) = \{|x - x_0| < c(t_0 - t)\}$, its boundary, i.e. the corresponding conic surface is $M_c^-(z_0) = \{|x - x_0| = c(t_0 - t)\}$. The symbols $B_R(z_0)$ $(S_R(z_0))$stand for the ball (sphere) centered at z_0 and with radius $R > 0$. Consider now the system

$$\begin{vmatrix} \Box_1 u = 0 \\ \Box_{1+\sigma} v = 0, \quad 0 \le \sigma \ll 1 \\ \Box_2 w = uv\psi(t,x), \quad \psi \in C^\infty. \end{vmatrix}$$

For $\sigma > 0$ the system is strictly hyperbolic and for $\sigma = 0$ – degenerate hyperbolic. The investigations in both cases are similar and we shall confine ourselves with the system

$$\begin{vmatrix} \Box_1 u = 0 \\ \Box_1 v = 0 \\ \Box_2 w = \psi uv \end{vmatrix} \tag{9.1}$$

equipped with Cauchy data on the non-characteristic hyperplane $\alpha : t = \frac{x_1 + x_2}{4}$ and such that $w|_{t < \frac{x_1 + x_2}{4}} = 0$, while the solutions u, v of $\Box_1 u = 0$,

$\Box_1 v = 0$ in $\mathbf{R}^3$ are $u = (t-x_1)^{p_1}\theta(t-x_1)$, $p_1 \in \mathbf{N}\cup\{0\}$, $v = (t-x_2)^{p_2}\theta(t-x_2)$, $p_2 \in \mathbf{N}\cup\{0\}$, $\psi \equiv \psi(4t-x_1-x_2)$, $\psi(z) = \tau^{p_3}\theta(\tau)$, $p_3 \in \mathbf{N}\cup\{0\}$. As usual, $\theta(\tau)$ is the standard Heaviside function.

This is the physical interpretation of our Cauchy problem (9.1). We are studying the propagation of 3 semilinear waves in 2D. Two of them are piecewise smooth travelling waves starting from $-\infty$ and the corresponding characteristics are $\Sigma_1 : t - x_1 = 0$, $\Sigma_2 : t - x_2 = 0$. Evidently, Σ_1 and Σ_2 are transversal to each other and $\Sigma_1 \cap \Sigma_2 = \Gamma$, $\alpha \cap \Sigma_1 \cap \Sigma_2 = 0$. The initial data of the third wave are prescribed on the non-characteristic plane α. Our waves collide at $\Gamma_+ = \Gamma \cap \{t > 0\}$. The hyperplanes Σ_1, Σ_2 are tangential to the characteristic cone of the future $M_1^+(0)$ but they are transversal to $M_2^+(0)$ and Γ_+ is located between $M_1^+(0)$ and $M_2^+(0)$. Due to the interaction of the waves at Γ_+ new singularities of w appear. One can show that w possesses square root – logarithmic type singularities and therefore w is not piecewise smooth.

The investigations here are based on appropriate changes of the variables, the finite speed of propagation of the waves satisfying $\Box_c w = 0$ and some elementary results from the distribution theory and the microlocal analysis. We shall discuss these singularities below.

As it is well known [85] a non-degenerate linear change of the variables is called a Lorentz one iff it conserves the operator $\Box_c$ up to the constant (velocity) $c > 0$, i.e. $(t, x) \to (\tau, y) \Rightarrow \Box_c \leadsto \Box_{\tilde{c}}$, $\tilde{c} > 0$. We are looking for a Lorentz transformation of the form

$$\begin{vmatrix} y_1 = \lambda(t-x_1) + (t-x_2) \\ y_2 = (t-x_1) + \lambda(t-x_2) \\ \tau = 4t - x_1 - x_2, \quad \lambda = \text{const} \neq 0. \end{vmatrix} \tag{9.2}$$

Thus, the Cauchy data for (9.1) are prescribed on the hyperplane $\tau = 0$ and we are looking for a solution in $\{\tau > 0\}$, $w|_{\tau<0} = 0$. We observe that $\operatorname{sing supp} u = \Sigma_1$, $\operatorname{sing supp} v = \Sigma_2$ and that the wedge $W = \{(t, x) : t \geq x_1, t \geq x_2\}$ has the edge $\Gamma : \{x_1 = x_2 = t\}$. If (9.2) is non degenerate, then (9.2) transforms W into the wedge $\tilde{W}$ whose edge $\tilde{\Gamma} = \{\tau = 2t, y_1 = 0, y_2 = 0\} = \overrightarrow{O\tau}$. Moreover, if $\lambda > 0$ then $\tilde{W}$ is the cartesian product of an acute central angle contained in $\mathbf{R}^2_{y_1,y_2}$ and $\overrightarrow{O\tau}$ (the positive axis $\{\tau \geq 0\}$). One can easily see that (9.2) is nondegenerate iff $\lambda^2 \neq 1$. Doing the corresponding computations we conclude that $\Box_2 w = 8\Box_c w$, when $c = 2 + \sqrt{2}$. So

$$\Box_c w = \frac{1}{8}\psi(\tau)u\left(\frac{\lambda y_1 - y_2}{\lambda^2 - 1}\right)v\left(\frac{\lambda y_2 - y_1}{\lambda^2 - 1}\right), \quad w|_{\tau<0} \tag{9.3}$$

2. We shall propose now a small excursion in the theory of fundamental solutions to second order constant coefficients hyperbolic equations. As it is obvious, $\frac{\partial^2}{\partial x_1 \partial x_2}\theta(x_1)\theta(x_2) = \delta(x_1)\delta(x_2)$. The non-degenerate linear change $x = Ay$, $A = \begin{pmatrix} k_1 & k_2 \\ k_3 & k_4 \end{pmatrix}$, $\det A \neq 0$ in the previous equation gives us that:

$$M_1 M_2 \left[\theta(k_1 y_1 + k_2 y_2)\theta(k_3 y_1 + k_4 y_2)\right] = \delta(Ay) = \frac{\delta(y)}{|\det A|}, \qquad (9.4)$$

where

$$M_1 = \frac{1}{\det A}\left(k_4 \frac{\partial}{\partial y_1} - k_3 \frac{\partial}{\partial y_2}\right), \quad M_2 = \frac{1}{\det A}\left(-k_2 \frac{\partial}{\partial y_1} + k_1 \frac{\partial}{\partial y_2}\right). \quad (9.5)$$

Geometrically we have that if $l_1 : k_1 y_1 + k_2 y_2 = 0$, $l_2 : k_3 y_1 + k_4 y_2 = 0$ are two lines in $\mathbf{R}^2_{y_1,y_2}$, then $M_1 \parallel l_2$ and $M_2 \parallel l_1$.

Let $\psi_1 \in D'(\mathbf{R})$, $\psi_2 \in D'(\mathbf{R})$. Then the following relations hold:

$$M_1(\psi_1(k_1 y_1 + k_2 y_2)) = \psi_1'(k_1 y_1 + k_2 y_2), \; M_1(\psi_2(k_3 y_1 + k_4 y_2)) = 0,$$
$$M_2\psi_1 = 0, \; M_2\psi_2 = \psi_2'(k_3 y_1 + k_4 y_2). \qquad (9.6)$$

For sufficiently smooth ψ_1, ψ_2 we have that

$$M_1 M_2(\psi_1 \psi_2) = M_1(\psi_2 M_2 \psi_1 + \psi_1 M_2 \psi_2) = M_1(\psi_1 M_2 \psi_2)$$
$$= M_1 \psi_1 M_2 \psi_2 + \psi_1 M_2 M_1 \psi_2 = M_1 \psi_1 M_2 \psi_2 \qquad (9.7)$$

as M_1, M_2 are commuting,

$$\ldots \quad M_1^p M_2^q(\psi_1 \psi_2) = M_1^p \psi_1 M_2^q \psi_2, p, q \in \mathbf{N}.$$

Taking in (9.3), (9.4) $k_1 = \frac{\lambda}{\lambda^2 - 1}$, $k_2 = \frac{-1}{\lambda^2 - 1}$, $k_3 = \frac{-1}{\lambda^2 - 1}$, $k_4 = \frac{\lambda}{\lambda^2 - 1}$, $\lambda = (1 + \sqrt{2})^2$, $u = \psi_1$, $v = \psi_2$ and applying $M_1^{p_1+1} M_2^{p_2+1}$ to (9.3) we get:

$$\Box_c M_1^{p_1+1} M_2^{p_2+1} w = \tfrac{1}{8}\psi(\tau) M_1 M_2 \left[(M^{p_1} u)(M^{p_2} v)\right],$$
$$M_1^{p_1+1} M_2^{p_2+1} w\Big|_{\tau<0} = 0. \qquad (9.8)$$

As we know,

$$M^{p_1} u M^{p_2} v = u^{(p_1)}\left(\frac{\lambda y_1 - y_2}{\lambda^2 - 1}\right) v^{(p_2)}\left(\frac{\lambda y_2 - y_1}{\lambda^2 - 1}\right)$$

$$= \tilde{c}_1 \theta\left(\frac{\lambda y_1 - y_2}{\lambda^2 - 1}\right)\theta\left(\frac{\lambda y_2 - y_1}{\lambda^2 - 1}\right)$$

and consequently (9.4) leads to

$$\Box_c M_1^{p_1+1} M_2^{p_2+1} w = \psi(\tau)\tilde{c}_2 \delta(y_1, y_2) \equiv g(\tau, y)$$
$$M_1^{p_1+1} M_2^{p_2+1} w\Big|_{\tau<0} = 0, \quad \tilde{c}_1, \tilde{c}_2 = \text{const} \neq 0. \qquad (9.9)$$

We have a formula [117], [56] for the generalized Cauchy problem giving the unique distribution solution of (9.9): $M_1^{p_1+1} M_2^{p_2+1} w = E_{2,c} * g(\tau, y)$ and $E_{2,c}$ is the fundamental solution $\Box_c E_{2,c} = \delta(\tau)\delta(y)$; $c = 2 + \sqrt{2}$, i.e. $E_{2,c} = \frac{1}{2\pi c} \frac{\theta(c\tau - |y|)}{\sqrt{c^2\tau^2 - |y|^2}}$. Then one can guess that

$$M_1^{p_1+1} M_2^{p_2+1} w(\tau, y) = \tilde{c}_3 \theta \left(\tau - \frac{|y|}{c} \right) \int_0^{\tau - |y|/c} \frac{\psi(\mu)d\mu}{\sqrt{(\mu - \tau)^2 - |y|^2/c^2}},$$

$$\tilde{c}_3 = \text{const} \neq 0.$$

$$(9.10)$$

On the other hand, $\psi(\mu) = \mu^{p_3}\theta(\mu)$. There are no difficulties to compute the integral in the right-hand side of (9.10) and to prove that the singularities that appear are of the type $\sqrt{}$ or $\log$ [94]. In the special case $p_3 = 0$ we have the following singularity: $\theta \left(\tau - \frac{|y|}{c} \right) \ln \frac{\frac{|y|}{c}}{\tau - \sqrt{\tau^2 - |y|^2/c^2}}$. In other words, a derivative of order $p_1 + p_2 + 2$ of w in the (τ, y) coordinates develop a logarithmic type singularity (not piecewise smooth) in the cone of the future $M_c^+(0)$ and along the ray $R_\tau^+ = \{\tau > 0, y = 0\}$. The solution is C^∞ smooth in $\left\{ 0 < \frac{|y|}{c} < \tau \right\}$, i.e. in $K_c^+(0) \backslash R_\tau^+$. In the (t, x) coordinates w develops $\log$ type singularities at $M_2^+(0)$ and Γ.

3. Our last observation concerns the case $p_1 = p_2 = 0$, $\psi \equiv 1 \Rightarrow \Box_2 w = uv$. Then in the new coordinates (τ, y) and without Cauchy data for (9.1), (9.3) the equation (9.9) takes the form

$$\Box_c M_1 M_2 w = \tilde{c}_2 \delta(y_1, y_2).$$

$$(9.11)$$

We are looking now for a solution w of (9.3), $\psi \equiv 1$, depending only on (y_1, y_2), i.e. $w(y)$ must satisfy

$$-c^2 \Delta_y w = \frac{1}{8} \theta \left(\frac{\lambda y_1 - y_2}{\lambda^2 - 1} \right) \theta \left(\frac{\lambda y_2 - y_1}{\lambda^2 - 1} \right).$$

$$(9.12)$$

In other words $M_1(\partial_y) M_2(\partial_y) w(y)$ is a fundamental solution of the Laplace operator in $\mathbf{R}_y^2$, i.e.

$$M_1 M_2 w = \tilde{c}_5 \ln |y| + C^\infty \text{ function}, \tilde{c}_5 \neq 0.$$

Going back to the coordinates (t, x) we conclude that $w(t, x)$ is a solution of $\Box_2 w = uv$ and it has a second derivative (more precisely $M_1(\partial_t, \partial_x) M_2(\partial_t, \partial_x) w$) developing a logarithmic singularity along the whole curve of intersection Γ of the incoming waves (extending to $-\infty$). In fact, $y = 0 \Leftrightarrow t - x_1 = 0$, $t - x_2 = 0 \Leftrightarrow t = x_1 = x_2$.

G. Métivier and J. Rauch ([80], [81]) proposed a modification of the above mentioned construction in order to find a solution w of the standard

Cauchy problem (9.1) with data at $\{t = 0\}$ and such that the interaction is localized. Their first step is to modify the third equation of (9.1). So we replace $\Box_2 w = uv$ by

$$\Box_2 \underline{w} = \eta(t, x)uv, \ \eta \in C_0^\infty(\mathbf{R}^3), \ 0 \le \eta \le 1,$$

$$\underline{w}(0, x) = \underline{w}_t(0, x) = 0; \ u = \theta(t - x_1), \ v = \theta(t - x_2).$$

(9.13)

Put $\overrightarrow{1} = (1, 1, 1) \in \Gamma$. Then we assume that $\operatorname{supp}\eta \subset B_{\frac{\varepsilon}{2\sqrt{2}}}(\overrightarrow{1})$ and $\eta \equiv 1$ on $B_{\frac{\varepsilon}{4}}(\overrightarrow{1})$, $0 < \varepsilon \ll 1$, ε – fixed. Having in mind that $(t, x) \in B_{\frac{\varepsilon}{2\sqrt{2}}}(\overrightarrow{1}) \Rightarrow |t - 1| < \frac{\varepsilon}{2\sqrt{2}}$ and therefore $1 - \frac{\varepsilon}{2\sqrt{2}} < t < 1 + \frac{\varepsilon}{2\sqrt{2}}$ on $\operatorname{supp}\eta$ we conclude that $\underline{w} = 0$ in the strip $0 \le t \le 1 - \frac{\varepsilon}{2\sqrt{2}}$. One can easily see that $B_{\frac{\varepsilon}{2\sqrt{2}}}(\overrightarrow{1}) \subset K_2^-(A)$, $A = \left(1 + \frac{\varepsilon}{2}, 1, 1\right)$ and that Γ is crossing $K_2^-(A)$ and $K_2^+(A)$ (see Fig. 9.1). Certainly, $\underline{w}$ does not vanish in $K_2^+(A)$ because of the finite speed of propagation of the waves.

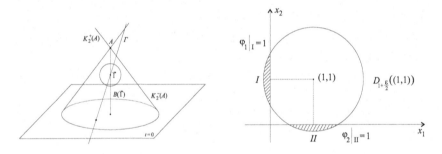

Fig. 9.1

We observe also that $B_{\frac{\varepsilon}{2\sqrt{2}}}(\overrightarrow{1}) \subset K_1^-(A) \subset K_2^-(A)$.

We shall work in the new coordinates (τ, y) investigating the wave fronts WF of w, u, v. We have just taken $w = w(y)$, while the change (9.2) implies that $u(t - x_1) = u\left(\frac{\lambda y_1 - y_2}{\lambda^2 - 1}\right) \equiv u(y)$ and $v(t - x_2) = v(y)$. Denote by (T, γ) the dual variables of (τ, y), $\operatorname{Char}\Box_c = \{(T, \gamma) : T^2 = c^2|\gamma|^2, (T, \gamma) \ne 0\}$ and $\operatorname{Ell}\Box_c = \{(T, \gamma) \ne 0 : T^2 \ne c^2|\gamma|^2\}$. Certainly, $\Box_c w = uv$ in $\mathbf{R}^3_{(\tau, y)}$. Formally, the Fourier transform $\widehat{w(y) \times 1(\tau)} = \hat{w}(\gamma)\delta(T)$, i.e. $WF(w) \subset \{T = 0\}$ and certainly, $WF(uv) \subset \{T = 0\}$. In fact $u(y)v(y) = (uv)(y)$. Therefore, $WF(w) \cup WF(uv) \subset \{T = 0\} \subset \operatorname{Ell}(\Box_c)$. The elementary facts from the microlocal analysis used here can be found in many books (see for example [56]). As WF and the characteristic set $\operatorname{Char}\Box_c$ are invariant under regular changes we obtain in the (t, x) coordinates the same inclusion, i.e.

$$WF(w) \cup WF(uv) \subset \text{Ell}(\square_2).$$

and consequently,

$$WF(w) \cap \text{Char}(\square_2) = \emptyset, \ WF(uv) \cap \text{Char}(\square_2) = \emptyset. \tag{9.14}$$

Having in mind that $WF(\eta uv) \subset WF(uv)$ as $\eta \in C_0^\infty$ we get that

$$WF(\eta uv) \cap \text{Char}(\square_2) = \emptyset. \tag{9.15}$$

Consider now the Cauchy problem $\square_2 \underline{w} = \eta uv$, $\underline{w}|_{t=0} = \underline{w}_t|_{t=0} = 0$. We know that $\underline{w} = 0$ for $0 \le t \le 1 - \frac{\varepsilon}{2\sqrt{2}}$. According to Theorem 26.1 from [56] – vol. IV, the set $WF(\underline{w})\backslash WF(\eta uv) \subset \text{Char}(\square_2)$ and is invariant under the flow defined by the Hamilton vector field $H_{\square_2}$. The integral curves $\mathcal{L}$ of $H_{\square_2}$ are generatrices of the characteristic conic surfaces and with equations $\mathcal{L}: t = 2\tau^0 s + t_0$, $x = -8\xi^0 s + x_0$, $\tau = \tau^0$, $\xi = \xi^0$, $(\tau^0, \xi^0) \in \text{Char}(\square_2)$. Obviously, $WF(\underline{w})\backslash WF(\eta uv) = \emptyset$ for $0 \le t \le 1 - \frac{\varepsilon}{2\sqrt{2}}$. Take a point B located in $\left\{t > 1 - \frac{\varepsilon}{2\sqrt{2}}\right\}$ and let $\mathcal{L}$ be the integral curve passing through B. Then $WF(\underline{w})\backslash WF(\eta uv) = \emptyset$ along $\mathcal{L}$. Thus, if $B \in WF(\underline{w}) \Rightarrow B \in WF(\eta uv)$ and consequently, $B \notin \text{Char}(\square_2)$ according to (9.15).

We conclude that

$$WF(\underline{w}) \cap \text{Char}(\square_2) = \emptyset. \tag{9.16}$$

Combining (9.14) and (9.16) we obtain

$$WF(w - \underline{w}) \cap \text{Char}(\square_2) = \emptyset \Rightarrow WF(w - \underline{w}) \subset \text{Ell}(\square_2). \tag{9.17}$$

In fact, if there exists $\rho \in WF(w - \underline{w})$, $\rho \in \text{Char}(\square_2)$ then (9.14), (9.16) imply that $\rho \notin WF(\underline{w})$, $\rho \notin WF(w)$ and by the definition of wave front set $\rho \notin WF(w - \underline{w})$ – contradiction. As we know, $\square_2(w - \underline{w}) = 0$ on $B_{\frac{\varepsilon}{4}}(\overrightarrow{1})$ as $\eta = 1$ there. The microlocal ellipticity of $\square_2$ on $\text{Ell}(\square_2)$ and (9.17) give us that $w - \underline{w} \in C^\infty$ on $B_{\frac{\varepsilon}{4}}(\overrightarrow{1})$. Thus we constructed a solution $\underline{w}$ of (9.13) which has logarithmic singularities near $\overrightarrow{1}$. As it is well known, $\frac{\partial u}{\partial t} = \delta'(t - x_1)$, $\frac{\partial v}{\partial t} = \delta'(t - x_2)$ and $WF(u)$, $WF(u_t) \subset \{(t = x_1, x_2; 1, -1, 0\}$, respectively $WF(v)$, $WF(v_t) \subset \{(t = x_2, x_1; 1, 0, -1\}$, i.e. $WF(u)$, $WF(v)$, $WF(u_t)$, $WF(v_t)$ are not intersecting the normal bundle over $\{t = 0\}$, namely $\{(t = 0, x; 1, 0, 0)\}$. Then according to [56] there exist the restrictions $u|_{t=0}$, $u_t|_{t=0}$, $v|_{t=0}$, $v_t|_{t=0}$. Consider now the following Cauchy problems

$$\square_1 \underline{u} = 0, \quad \square_1 \underline{v} = 0, \tag{9.18}$$

$$\underline{u}(0,x) = \varphi_1(x)u(0,x), \quad \underline{u}_t(0,x) = \varphi_1(x)u_t(0,x) \tag{9.19}$$

$$\underline{v}(0,x) = \varphi_2(x)v(0,x), \quad \underline{v}_t(0,x) = \varphi_2(x)v_t(0,x) \tag{9.20}$$

and $\varphi_{1,2} \in C_0^\infty(\mathbf{R}^2)$ will be chosen later. Evidently, supp u, $u_t \subset \{x_1 - t \leq 0\} \Rightarrow$ supp $u|_{t=0}$, $u_t|_{t=0} \subset \{(x_1,x_2) : x_1 \leq 0\}$ and supp $v|_{t=0}$, $v_t|_{t=0} \subset \{(x_1,x_2) : x_2 \leq 0\}$. Denote by $D_R(Q)$ the disc in $\mathbf{R}_x^2$ of radius R and centered at Q. Then $K_1^-(A) \cap \{t = 0\} = D_{1+\frac{\varepsilon}{2}}((1,1))$ (see Fig. 9.1) and $supp(u|_{t=0}, u_t|_{t=0}) \cap D_{1+\frac{\varepsilon}{2}}((1,1)) \subset I$ (I being the segment of the disc contained in $\{x_1 \leq 0\}$), supp $(v|_{t=0}, v_t|_{t=0}) \cap D_{1+\frac{\varepsilon}{2}}((1,1)) \subset II$ (II being the segment of the disc contained in $\{x_2 \leq 0\}$). We take now the cutoff function φ_1 with a support in a tiny neighbourhood of I, $\varphi_1 = 1$ near I and supp φ_2 is in a tiny neighbourhood of II, $\varphi_2 = 1$ near II. Thus, the Cauchy data for u, v, respectively $\underline{u}$, $\underline{v}$, coincide in $D_{1+\frac{\varepsilon}{2}}((1,1)) \Rightarrow u = \underline{u}$, $v = \underline{v}$ in $K_1^-(A) \Rightarrow uv = \underline{u}\,\underline{v}$ in $K_1^-(A) \Rightarrow \eta uv = \eta\underline{u}\,\underline{v}$ everywhere as $supp\,\eta \subset K_1^-(A)$. Thus, $\underline{w}$ satisfies

$$\Box_2 \underline{w} = \eta\underline{u}\,\underline{v}$$
$$\underline{w}|_{t=0} = \underline{w}_t|_{t=0} = 0. \tag{9.21}$$

Proposition 9.1. [80] *The Cauchy problem (9.18), (9.19), (9.20), (9.21) of 3 equations is globally solvable and $\underline{w}$ possesses a second derivative with logarithmic singularity along Γ_+ (at least near $\overrightarrow{1}$). In other words, failure of piecewise smoothness occurs uniquely along the line of interaction Γ_+.*

Let us add a fourth equation transporting the logarithmic singularities of $\underline{w}$ along a hyperplane containing Γ_+. Thus,

$$\partial_t z = \zeta(t,x)\underline{w}$$
$$z(0,x) = 0, \tag{9.22}$$

$\zeta \in C_0^\infty$, supp $\zeta \subset \{(t,x) : \eta(t,x) = 1\}$ and $\zeta \equiv 1$ in a tiny neighbourhood of $\overrightarrow{1}$. Make at first the translation $(t,x) \to (t',x') = (t,x) - \overrightarrow{1}$. Consider then the change of the variables $(t',x') \to (T,X)$:

$$t' = X_2 + T$$
$$x_1' = X_1 + X_2 \tag{9.23}$$
$$x_2' = 2X_1 + X_2.$$

One can easily guess that; $\overrightarrow{1} \to (0,0,0)$, $\frac{\partial}{\partial T} = \frac{\partial}{\partial t'} = \frac{\partial}{\partial t}$, the line Γ is mapped onto the line $X_1 = T = 0$, the hyperplane $x_1' = x_2'$ is mapped onto $X_1 = 0$ and that the plane $t' = 0$ is not mapped onto the plane $T = 0$ (see Fig. 9.2).

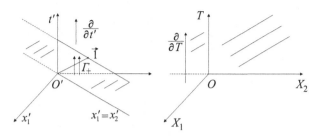

Fig. 9.2

As we know from (9.2): $y = 0 \Leftrightarrow x_1' = x_2' = t' \Leftrightarrow X_1 = T = 0$, i.e. $0 < \frac{|y|^2}{T^2 + X_1^2} \in C^\infty$ near the origin. Therefore, in a neighbourhood of O in $\mathbf{R}^3_{TX_1X_2}$ we have that $\zeta = 1$ and $\frac{\partial}{\partial T} z = \underline{w} \Rightarrow \frac{\partial}{\partial T} M_1 M_2 z = M_1 M_2 \underline{w} = \tilde{c}_5 \ln(T^2 + X_1^2) + C^\infty$. Certainly, M_1, M_2 are written in the (T, X) coordinates. So $M_1 M_2 z = \tilde{c}_5 \int_0^T \ln(T^2 + X_1^2) dT + f(X) + C^\infty$, $f(X) \in D'(\mathbf{R}^2_X) \Rightarrow$
$$M_1 M_2 z = \tilde{c}_5 \left(2X_1 \arctan \frac{T}{X_1} + T \ln(T^2 + X_1^2) - 2T \right) + f(X) + C^\infty \equiv$$
$h(T, X) + f(X) + C^\infty$.

Evidently, $\frac{\partial^2 h}{\partial X_1^2} = -\frac{2T\tilde{c}_5}{X_1^2 + T^2} \Rightarrow \lim_{T \to 0} \left| \frac{\partial^2 h}{\partial X_1^2}(T, 0) \right| = +\infty$. The rate of explosion $-\frac{1}{T}$ depends on T and cannot be cancelled by $F(X)$. So we conclude that a 4th derivative of z has logarithmic singularity at the hyperplane $x_1 = x_2$ which is not spacelike for $\Box_{1,2}$ (see that definition in [15]).

Proposition 9.2. [80] *The hyperbolic system (9.18)–(9.22) of 4 equations possesses logarithmic singularities produced by the interaction of 2 piecewise smooth waves along the straight line Γ_+ which is not contained in a spacelike manifold. The failure of piecewise smoothness for one of the solutions $\underline{w}$ occurs uniquely along the line of interaction, while the second one z has logarithmic singularities along a hyperplane containing Γ_+.*

In fact, the logarithmic singularities of $\underline{w}$ at Γ_+ propagate as logarithmic singularities to z along the integral curves of $\frac{\partial}{\partial t}$ contained in the hyperplane $x_1 = x_2$ (see Fig. 9.2).

The fact that in the $2D$ case logarithmic singularities appear for the wave equation was observed many years ago in [36]. Similar result is proved in [68].

Bibliography

[1] Abdelouhab, L., Bona, J., Felland, M., and Saut, J. (1989). Nonlocal models for nonlinear, dispersive waves, *Phys. D*, vol. 40, pp. 360–392.

[2] Ahiezer, N. (1970). *Elements of the theory of elliptic functions*, Edition "Nauka", Moscow (in Russian).

[3] Albert, J. and Bona, J. (1991). Total positivity and the stability of internal waves in fluids of finite depth, *IMA J. Appl. Math.*, vol. 46, pp. 1–19.

[4] Albert, J. (1992). Positivity properties and stablilty of solitary wave solutions of model equations for long waves, *Comm. Part. Diff. Equations*, vol. 17, pp. 1–22.

[5] Angulo, J. (2007). Non-linear stability of periodic travelling wave equation for the Schrödinger and modified Korteweg-de Vries equation, *J. Differ. equations*, vol. 235, pp. 1–30.

[6] Angulo, J. and Natali, F. (2008). Positivity properties of the Fourier transform and the stability of periodic travelling wave solutions, *SIAM J. Math. Anal.*, vol. 40:3, pp. 1123–1151.

[7] Angulo, J., Bona, J. and Scialom, M. (2006). Stability of cnoidal waves, *Adv. Diff. Equations*, vol. 11, pp. 1321–1374.

[8] Antosik, P., Mikusinski, J. and Sikorski, R. (1973). *Theory of distributions. The sequential approach.* Elsevier, Amsterdam.

[9] Arhangel'skii, J. (1977). *Solid body analytic dynamics*, Edition "Nauka", Moscow, (in Russian).

[10] Arnol'd, V. (1971). *Ordinary differential equations*, Edition "Nauka", Moscow, (in Russian).

[11] Beals, R., Sattinger, D. and Szmigielski, J. (1999). Multipeakons and a theorem of Stieltjes, *Inverse problems*, vol. 15, Letters L1–L4.

[12] Beals, R., Sattinger, D. and Szmigielski, J. (2000). Multipeakons and the classical moment problem, *Adv. Math.*, vol. 154, pp. 229–257.

[13] Benjamin, T. (1972). The stability of solitary waves, Proc. Royal Soc. London Ser. A, 338, 153-183.

[14] Benjamin, T. (1996). Solitary and periodic waves of a new kind, *Philos. Trans. Roy. Soc. London* Ser. A, vol. 354, pp. 1775–1806.

[15] Bers, L., John, F. and Schechter, M. (1964). *Partial differential equations*, Interscience publishers, NY-London-Sydney.

[16] Bhatnagar, P. (1979). *Nonlinear waves in dimensional dispersive systems*, Oxford monograph, Clarendon Press Oxford.

[17] Bona, J. (1975). On the stability theory of solitary waves, *Proc. Royal Soc. London*, Ser. A, vol. 344, pp. 363–374.

[18] Bressan, A. and Constantin, A. (2005). Global solutions of the Hunter-Saxton equation, *SIAM J. Math. Anal.*, vol. 37, pp. 996–1026.

[19] Bressan, A. and Constantin, A. (2007). Global conservative solutions of the Camassa-Holm equation, *Arch. Rational Mech. Anal.*, vol. 183, pp. 215–239.

[20] Bressan, A. and Constantin, A. (2007). Global dissipative solutions of the Camassa-Holm equation, *Analysis and Appl.*, vol. 5:1, pp. 1–27.

[21] Byrd, P. and Friedman, M. (1971). *Handbook of elliptic integrals for engineers and scientists*, Springer-Verlag, NY.

[22] Camacho, V., Guy, R. and Jacobsen, J. (2008). Travelling waves and shocks in a viscoelastic generalization of Burger's equation, *SIAM J. Appl. Math*, vol. 68:5, pp. 1316–1332.

[23] Camassa, R. and Holm, D. (1993). An integrable shallow water equation with peaked solutions, *Phys. Rev. Letters*, vol. 71, pp. 1661–1664.

[24] Camassa, R., Holm, D. and Hyman, J. (1994). A new integrable shallow water equation, *Adv. Appl. Mech.*, vol. 31, pp. 1–33.

[25] Chua, L. and Yang, L. (1988). Cellular Neural Network: Theory and Applications, *IEEE Trans. CAS*, vol. 35, pp. 1257–1290.

[26] Chua, L., Hasler, M., Moschytz, G. and Neirynsk, J. (1995). Autonomous cellular neural networks: a unified paradigm for pattern formation and active wave propagation, *IEEE Trans. CAS-I*, vol. 42:10, pp. 559–577.

[27] Colliander, J., Keel, M., Stafilani, G., Takaoka, H. and Tao, T. (2003). Sharp global well-posedness for the KdV and modified KdV on $\mathbb{R}^1$ and $\mathbb{T}$, *J. Amer. Math. Soc.*, vol. 16, pp. 705–749.

[28] Colombeau, J. (1990). Multiplication of distributions, *Bull. Amer. Math. Soc.*, vol. 23:2, pp. 251–268.

[29] Constantin, A. and Escher, J. (1998). Global existence and blow up for a shallow water equation, *Ann. Sc. Norm. Sup. Pisa*, IV Ser., vol. 26, pp. 303–328.

[30] Constantin, A. and Escher, J. (1998). Wave breaking for nonlinear nonlocal shallow water equations, *Acta Math.*, vol. 181, pp. 229–243.

[31] Constantin, A. (2000). Existence of permanent and breaking waves for a shallow water equation: A geometric approach, *Ann. Inst. Fourier (Grenoble)*, vol. 50, pp. 321–362.

[32] Constantin, A. and Molinet, L. (2000). Global weak solutions for a shallow water equation, *Comm. Math. Phys.*, vol. 211, pp. 45–61.

[33] Constantin, A. and Ivanov, R. (2008). On an integrable two-component Camassa-Holm shallow water system, *Physics Letters A*, www.elsevier.com/locate/pla, doi : 10 : 1016/j.physlet.a.2008.10.050.

[34] Constantin, A. and Strauss, W. (2000). Stability of peakons, *Comm. on Pure and Appl. Math.*, vol. 53, pp. 603–610.

[35] Cooper, F., Shepard H. and Sodano, P. (1993). Solitary waves in a class of generalized Korteweg-de Vries equation, *Phys. Rev. E*, vol. 48:5, pp. 4027–32.

[36] Courant, R. (1962). *Partial differential equations II*, Interscience Publishers, NY.

[37] Courant, R. and Friedrichs, K. (1976). Supersonic flow and shock waves, *Appl. Math. Series*, vol. 21, Springer Verlag, NY-Heidelberg-Berlin.

[38] Dafermos, C. (1977). Generalized characteristics and the structure of solutions of hyperbolic conservation laws, *Indiana U. Math. J.*, vol. 26, pp. 1097–1119.

[39] Dafermos, C. (2005). *Hyperbolic conservation laws in continuum physics* (Second edition), Springer.

[40] Dafermos, C. (2006). Continuous solutions for balance laws, *Ricerche di Matematica*, vol. 55, pp. 79–91.

[41] Danilov, V. and Shelkovich, M. (2005). Delta-shock wave type solution of hyperbolic systems of conservation laws, *Quart. Appl. Math.*, vol. 63:3, pp. 401–427.

[42] Danilov, V. and Shelkovich, M. (2005). Dynamics of propagation and interaction of delta-shock waves in conservation law systems, *J. Diff. Equations*, vol. 211, pp. 333–381.

[43] Degasperis, A. and Procesi, M. (1999). *Asymptotic integrability, symmetry and perturbation theory*, in: Degasperis, A. and Gaeta G. (Eds), World Scientific, Singapore, pp. 23–37.

[44] Dullin, H., Gottwald, G. and Holm, D. (2001). An integrable shallow water equation with linear and non-linear dispersion, *Phys. Rev. Lett.*, vol. 87:19, 194501-1-4.

[45] Dwight, H. (1961). *Tables of integrals and other mathematical data*, Fourth edition, The Macmillan company, NY.

[46] El Dika, K. and Molinet, L. (2009). Stability of multi antipeakon-peakons profile, *Discrete and Cont. Dynam. Systems*, Series B, vol. 12:3, pp. 561–577.

[47] Evans, L. (1998). *Partial differential equations*, Amer. Math. Soc., Providence, RI.

[48] Fihtengholtz, A. (1962). *A course of differential and integral calculus*, vol. 2, 3, Edition Fizmatgiz, Moscow (in Russian).

[49] Fokas, A. and Fuchssteiner, B. (1981) Symplectic structures, their Backlund transformation and hereditary symmetries, *Physica D*, vol. 4, pp. 47–66.

[50] Forsyth, A. (1959). *Theory of differential equations*, vol. 6, Dover Publications, NY.

[51] Fortuna, L., Frasca, M. and Rizzo, A. (2002). Self-organizing behavior of arrays of non identical Josephson junctions, *2002 IEEE International Symposium on curcuits and systems*, vol. 5, pp. 213–216.

[52] Gel'fand, I. and Shilov, G. (1964). *Generalized functions I, Properties and operations*, Academic Press, NY.

[53] Grillakis, M., Shatah, J. and Strauss, W. (1987). Stability theory of solitary waves in the presence of symmetry, *J. Funct. Anal.*, vol. 74, pp. 160–197.

[54] Grillakis, M., Shatah, J. and Strauss, W. (1990). Stability theory of solitary waves in the presence of symmetry, *J. Funct. Anal.*, vol. 94, pp. 308–348.

[55] He Bin, Li Jibin, Long Yao and Weiguo, R. (2008). Bifurcations of travelling wave solutions for a variant of Camassa-Holm equation, *Nonlinear Analysis: Real World Appl.*, vol. 9, pp. 222–232.

[56] Hörmander, L. (1985). *The Analysis of linear partial differential operators*, vol. I-IV, Springer Verlag, NY.

[57] Hsu, C. and Lin, S. (1998). Travelling waves in lattice dynamical systems with applications to cellular neural networks, preprint.

[58] Hunter, J. and Saxton, R. (1991). Dynamics of director fields, *SIAM J. Appl. Math*, vol. 31:6, pp. 1498–1521.

[59] John, F. (1989). *Non-linear wave equations. Formation of singularities*, Univ. Lecture series, Amer. Math. Soc., Providence RI.

[60] Joseph, K. (1993). *Asymptotic analysis*, vol. 7, pp. 105–120.

[61] Karlin, S. (1968). *Total positivity*, Stanford University Press, Stanford, California.

[62] Khare, A. and Cooper, F. (1993). One parameter family of solutions with compact support in a class of generalized Korteweg-de Vries equation, *Phys. Rev. E*, vol. 58:6, pp. 4843–4.

[63] Knobel, R. (2000). *An introduction to the mathematical theory of waves*, Student Math. Library, vol. 3, American Math. Soc.

[64] Kraenkel, R. and Zenchuk, A. (1999). A Camassa-Holm equation: transformation to deformed sinh-Gordon equations, cuspon and soliton solutions, *J. Phys. A: Math. Gen*, vol. 32, pp. 4733–4747.

[65] Kranzer, H. and Keyfitz, B. (1990). A strictly hyperbolic system of conservation laws admitting singular shocks, in *Nonlinear evolution equations that change type*, IMA vol. Math. Applications, 27, Springer, NY, pp. 107–125.

[66] Kruzhkov, S. (1970). First order quasilinear equations with several space variables, Math. Sbornik, vol. 123, pp. 228–255.

[67] Lamb, G. (1971). Analytical descriptions of ultrashort optical pulse propagation in a resonant medium, *Rev. Mod. Phys.*, vol. 43, pp. 99–124.

[68] Laschon, G. (2000). Creation and propagation of two piecewise smooth progressing waves, *Proceedings of the Amer. Math. Soc.*, vol. 129, pp. 1375–1384.

[69] Lax, P. (1957). Hyperbolic systems of conservation laws II, *Comm. Pure Appl. Math.*, vol. 10, pp. 537–566.

[70] Lax, P. (1968). Integrals of nonlinear evolution equations and solitary waves, *Comm. Pure Appl. Math.*, vol. 21, pp. 467–490.

[71] Lax, P. (1973). Hyperbolic systems of conservation laws and the mathematical theory of shock waves, *CB MS- NSF Regional Conf. Ser. Appl. Math.*, vol. 11, SIAM, Philadelphia, PA.

[72] Lax, P. (2006). Hyperbolic Partial Differential Equations, *Courant Lecture Notes in Math.*, vol. 14, Amer. Math. Soc.

[73] Li Ji-bin (2008). Exact travelling wave solutions to 2D generalized Benney-Luke equation, *Appl. Math. Mech.*, vol. 29:11, pp. 1391–1398.

[74] Lin, Z. and Liu, Y. (2009). Stability of peakons for the Degasperis-Procesi equation, *Comm. Pure Appl. Math.*, vol. 62, pp. 125–146.

[75] Lin, Z. and Qian, T. (2002). Peakons of the Camassa-Holm equation, *Appl. Math. Modeling*, vol. 26, pp. 473–480.

[76] Lonngren, K. and Scott, A. (editors) (1978). Solitons in action, *Proc. of a workshop at Redstone arsenal*, Academic Press, NY.

[77] Magnus, W. and Oberhettinger, E. (1986). *Formulas and theorems for special functions of mathematical physics*, Springer, NY.

[78] Mallet-Paret, J. (1996). Spatial patterns, spatial chaos and travelling waves in lattice of differential equations, in: *Stochastic and Spatial Structure of Dynamical Systems*, editors van Strien, S. and Verduyn Lunel, S., North-Holland, pp. 105–129.

[79] McKean, H. (1997). Stability for the Korteweg-de Vries equation, *Comm. Pure Appl. Math.*, vol. 30, pp. 347–352.

[80] Métivier, G. and Rauch, J. (1989). The interaction of two progressive waves, *Lecture Notes in Math.*, vol. 1402, NY, Springer, pp. 216–226.

[81] Métivier, G. and Rauch, J. (1990). Interaction of piecewise smooth progressing waves for semilinear hyperbolic equations, *Communications in PDE*, vol. 15, pp. 239–289.

[82] Molinet, L. (2007). Global well-posedness in the energy space for the Benjamin-Ono equation on the circle, *Math. Annalen*, vol. 337, pp. 353–383.

[83] Molinet, L. (2008). Global well-posedness in L^2 for the periodic Benjamin-Ono equation, *Amer. J. Math.*, vol. 130, pp. 635–683.

[84] Molinet, L. and Riband, P. (2006). Well-posedness in H^1 for the (generalized) Benjamin-Ono equation on the circle, preprint.

[85] Naimark, M. (1958). *Linear representation of the Lorentz group*, Fizmatgiz, Moscow (in Russian).

[86] Oberguggenberger, M. (1992). Multiplication of distributions and applications to partial differential equations, *Pitman Res. Notes Math. Ser.*, vol. 259, Longman, Harlow; Wiley NY.

[87] Oberhettinger, E. (1973). *Fourier transforms of distributions and their inverses*, Academic Press, NY.

[88] Oleinik, O. (1957). Discontinuous solutions of non-linear differential equations, *Soviet Math. Surveys*, vol. 12, pp. 3–73.

[89] Parker, A. (2007). Cusped solutions of the Camassa-Holm equation I. Cuspon solitary wave and antipeakon limit, *Chaos, solitons and fractals*, vol. 34, pp. 730–739.

[90] Perko, L. (1991). *Differential equations and dynamical systems*, Springer, NY.

[91] Perring, J. and Skyrme, T. (1962). A model of unified field equation, *Nucl. Phys.*, vol. 31, pp. 550–555.

[92] Polijanin, A., Zaitcev, V. and Jurov, A. (2005). *Methods for solving of non-*

linear equations of mathematical physics and mechanics, Fizmatlit, Moscow (in Russian).

[93] Popivanov, P. (2003). Nonlinear PDE. Singularities, Propagation, Applications, *Birkhäuser series "Operator theory, Advances and Applications"*, vol. 145, pp. 1–94.

[94] Popivanov, P. (2006). Geometrical methods for solving of fully nonlinear partial differential equations, *Math. and its Appl. Series*, vol. 2, Union of Bulgarian Math.

[95] Popivanov, P. and Slavova, A. (2009). Peakons, cuspons, compactons, solitons, kinks and periodic solutions of several third order PDE and their CNN realization, *Lecture Notes in Computer Science*, vol. 5434, pp. 461–468.

[96] Popivanov, P., Slavova, A. and Zanghirati, L. (2009). Travelling waves for several equations of mathematical physics. Existence and profiles, *C. R. Acad. Bulg. Sci.*, vol. 62:3, pp. 303–314.

[97] Popivanov, P., Slavova, A. and Zecca P. (2009). Compact travelling waves and peakon type solutions of several equations of mathematical physics and their CNN realization, *Nonlinear Analysis: Real World Appl.*, vol. 10, pp. 1453–1465.

[98] Rauch, J. and Reed, M. (1980). Propagation of singularities for semilinear hyperbolic equations in one space variable, *Annals of Math.*, vol.111:2, pp. 531–552.

[99] Rauch, J. and Reed, M. (1982). Propagation of singularities in non-strictly hyperbolic semilinear systems; examples, *Comm. Pure Appl. Math.*, vol. 35, pp. 555–565.

[100] Rijik, I. and Gradstein, M. (1962). *Handbook on integrals, sums, series and products*, Edition Fizmatgiz, Moscow (in Russian).

[101] Rosenau, P. (2005). What is ... a compacton?, *Notices of the AMS*, vol. 52, pp. 738-739.

[102] Rosenau, P. (2006). On a model equation of travelling and stationary compactons, *Phys. Lett. A*, vol. 356, pp. 44–50.

[103] Rosenau, P. and Hyman, J. (1993). Compactons: solitons with finite wave length, *Phys. Rev. Lett.*, vol. 70, pp. 564–567.

[104] Roska, T., Chua, L., Wolf, D., Kozek, T., Tetzlaff, R. and Puffer, F. (1995). Simulating nonlinear waves and partial differential equations via CNN – Part I: Basic techniques, *IEEE Trans. CAS-I*, vol. 42:10, pp. 807–815.

[105] Shatah, J. and Struwe, M. (1998). Geometric wave equations, *Courant lecture notes in Math.*, vol. 2, Courant Inst. of Math. Sci., AMS, Providence, RI.

[106] Shelkovich, V. (2006). The Riemann problem admitting δ, δ'–shocks, and vacuum states (the vanishing viscosity approach), *J. Differ. Equations*, vol. 231, pp. 459–500.

[107] Shelkovich, V. (2008). δ and δ' shock waves of conservation law systems and transfer and concentration processes, *Russian Math. Surveys*, vol. 63:3, pp. 73–146.

[108] Shen, J. and Xu, W. (2007). Travelling wave solutions in a class of gener-

alized Korteweg-de Vries equations, *Chaos, solitons and fractals*, vol. 34, pp. 1299–1306.

[109] Slavova, A. (2003). *Cellular Neural Networks: Dynamics and Modelling*, Kluwer Academic Press.

[110] Stein, E. (1970). *Singular integrals and differentiability properties of functions*, Princeton University Press, Princeton, NJ.

[111] Stein, E. and Weiss, G. (1970). *Introduction to Fourier Analysis on Euclidean spaces*, Princeton University Press, Princeton, NJ.

[112] Tang, M. and Zhang, W. (2007). Four types of bounded solutions of CH- γ equation, *Science in China Series A:Mathematics*, vol. 50:1, pp. 132–152.

[113] Tao, T. (2009). Why are solitons stable? *Bull. Amer. Math. Soc.*, vol. 46, pp. 1–33.

[114] Thiran P., Crounse, K., Chua, L. and Hasler M. (1995). Pattern formation properties of autonomous cellular neural networks, *IEEE Tran. CAS*, vol. 42, pp. 754–774.

[115] Trees, B. and Stround, D. (1999). Two-dimensional arrays of Josephson junctions in a magnetic field: a stability analysis of synchronised states, *Physical Review B*, vol. 59:10, pp. 7108–7115.

[116] Tsuji, M. and Yamamoto M. (2008). Another approach to δ-shocks, *C. R. Acad. Bulg. Sci.*, vol. 61:6, pp. 701–704.

[117] Vladimirov, V. (1976). *Generalized Functions in Mathematical Physics*, Nauka, Moscow (in Russian).

[118] Weinstein, M. (1986). Lyapunov stability of ground states of nonlinear dispersive evolution equations, *Comm. Pure Appl. Math.*, vol. 39, pp. 51–67.

[119] Whitham, G. (1974). *Linear and nonlinear waves*, J. Wiley & Sons.

[120] Xin, Z. and Zhang, P. (2000). On the weak solutions to a shallow water equation, *Communications Pure Appl. Math.*, vol. 53, pp. 1411–1433.

[121] Yin, Z. (2004). On the blow up scenario for the generalized Camassa-Holm equation, *Comm. in PDE*, vol. 29:5, 6, pp. 867–877.

[122] Zhang, P. and Zheng, Y. (2005). On the global weak solutions to a variational wave equation, in: *Handbook of Differential Equations, Evolutionary equations*, vol. 2, Dafermos, C. and Feireisl editors, Elsevier, Amsterdam, pp. 561–648.

Index